Clécio Danilo Dias-da-Silva

Teaching Natural Sciences

Clécio Danilo Dias-da-Silva

Teaching Natural Sciences

Research, dialogue and reflection

ScienciaScripts

Imprint

Any brand names and product names mentioned in this book are subject to trademark, brand or patent protection and are trademarks or registered trademarks of their respective holders. The use of brand names, product names, common names, trade names, product descriptions etc. even without a particular marking in this work is in no way to be construed to mean that such names may be regarded as unrestricted in respect of trademark and brand protection legislation and could thus be used by anyone.

Cover image: www.ingimage.com

This book is a translation from the original published under ISBN 978-613-9-64962-4.

Publisher:
Sciencia Scripts
is a trademark of
Dodo Books Indian Ocean Ltd. and OmniScriptum S.R.L publishing group

120 High Road, East Finchley, London, N2 9ED, United Kingdom
Str. Armeneasca 28/1, office 1, Chisinau MD-2012, Republic of Moldova, Europe
Printed at: see last page
ISBN: 978-620-7-84585-9

ABOUT THE AUTHORS:

CLÉCIO DANILO DIAS DA SILVA

She has a degree in Biological Sciences from the FACEX University Centre - UNIFACEX (2015). She specialises in Teaching Natural Sciences and Mathematics at the Federal Institute of Education, Science and Technology of Rio Grande do Norte - IFRN (2017). Master's in Teaching Natural Sciences and Maths from the Federal University of Rio Grande do Norte - PPGECNM/ UFRN (2018). He is currently working at the UNIFACEX University Centre as a collaborating researcher on Scientific Initiation Research projects (PROIC).

GLAUBER HENRIQUE BORGES DE OLIVEIRA SOUTO

She has a degree in Biological Sciences from Universidade Potiguar - UnP (2007). She has a Master's degree in Biological Sciences from the Federal University of Rio Grande do Norte - UFRN (2010). She is currently working on her undergraduate degree in Biological Sciences and her postgraduate degree in Microbiology and Parasitology at the Facex University Centre (UNIFACEX).

LÚIS GUILHERME BEZERRA MACIEL

She has a degree in Biological Sciences from the FACEX University Centre - UNIFACEX (2015).

NAAMA PEGADO FERREIRA

She has a degree in Biological Sciences from the Federal University of Rio Grande do Norte - UFRN (2012). She specialised in Teaching Natural Sciences and Mathematics at the Federal Institute of Education, Science and Technology of Rio Grande do Norte - IFRN (2017). She is currently studying for a Master's degree in Teaching Natural Sciences and Maths at the Federal University of Rio Grande do Norte (PPGECNM/UFRN) and is a permanent teacher in the State of Rio Grande do Norte (SEEC/RN).

SUMMARY

PRESENTATION

It is with great satisfaction that I present in this book part of the academic experiences of various professionals in the classroom, involving themes from the natural sciences.

This material contains the results of research in the educational context, involving different themes within the Natural Sciences area, such as: botany, zoology, environmental education, microbiology, human physiology, among others. These materials provide dialogues and reflections on the implementation of various elements of research such as: application and validation of teaching sequences/units, analysis of student perceptions, mapping of scientific productions, bringing contributions to the implementation of investigations involving the teaching of Natural Sciences.

We hope that this book will serve as an incentive to improve your future classes, making them ever more dynamic, participatory and creative. May you feel encouraged not only to learn, but to share these and other didactic experiences with us in favour of quality education for our country, training scientifically literate students.

The organiser

CHAPTER 1

BIRDWATCHING AS A METHODOLOGICAL TOOL IN ENVIRONMENTAL EDUCATION

Luís Guilherme Bezerra Maciel[1,2]

Clécio Danilo Dias da Silva Glauber[2]

Henrique Borges de Oliveira Souto[3]

INTRODUCTION

Brazil is considered a Biologically Healthy Nation. This designation refers to its great diversity of ecosystems and species of flora and fauna. Among the representatives of fauna, birds are one of the best known and most studied groups in Brazil. Due to the vast knowledge of this group, birds have numerous functions, both for the maintenance of ecosystems and for man. These functions are: pest controllers, pollinators, seed dispersers and bioindicators of environmental quality.

Because birds are so beautiful, have attractive songs and arouse great fascination, it is very common for groups of birdwatchers to form. Bird watching, or birdwatching as the activity is also known, is all about spotting animals in their natural environments, collecting records using binoculars or photographic equipment when possible. It is a rewarding alternative leisure activity, as it develops the intellect and scientific knowledge directly in natural environments, or even in the city. It is a very common activity in the south and southeast of Brazil, and there are around 80 million birdwatchers worldwide. Birdwatching has great potential for teaching formal content that is usually dealt with in the classroom, serving as an effective tool for Environmental Education (EE).

According to Allenspach and Zuin (2013), between 1990 and 2012, 49 environmental education projects were registered using the theme of birds as a subsidy, not all of them

---

1 Degree in Biological Sciences (UNIFACEX)
2 Master in Teaching Natural Sciences and Mathematics (UFRN).
3 Master in Biological Sciences (UFRN).

focussed on observation. These projects are developed for a wide range of audiences and are carried out in the south and southeast, demonstrating the lack of them in the northeast.

Vieira-da-Rocha and Molin (2008) used birdwatching only in the classroom, combining it with other subjects, and demonstrated that the method was very effective, arousing great interest among students in the innovative method. Fieker (2012) and Lopes and Santos (2004) showed that birdwatching in the natural environment is very effective as an ecotourism and environmental education activity, arousing interest and satisfaction among participants.

The use of environmental education (EE) has become one of the most skilful and efficient ways of achieving better conservation of natural resources. This is because it makes it possible to raise individual or collective awareness of environmental problems and causes among the human population, making it an effective approach for those living on the margins of environmental protection zones.

With this in mind, the aim of this work is to arouse and stimulate the interest of students at the José Fernandes Machado State School by observing birds in the school grounds, thus demonstrating the interactions of these species in natural environments and using their previous knowledge as a strategy for environmental education, so that they can conserve and even preserve the species that live there and in other areas.

THEORETICAL REFERENCE

BRAZIL'S BIODIVERSITY

Brazil is home to an immense biological diversity, due to its territorial extension combined with its geographical location and climate, thus presenting a great variety of biomes, from the humid tropics in the north, the semi-arid in the northeast and the temperate regions in the south, thus making it a megadiverse country, possessing approximately 20% of the species of organisms ever described (PEREIRA; SILVA, 2009).

Birds are considered to be the most diverse group of terrestrial vertebrates, as they are widely distributed across the globe and have the greatest diversity of habits, habitats and structural variety (POUGH, 2003). The main characteristic of this group is the presence of feathers covering their bodies, which makes them easy to identify, but they are also characterised by their beaks, which can tell a lot about the species' eating habits and consequently their habitat (EFE, 1999). In addition to their great beauty and vocalisations that are often attractive to humans, they are considered to be animals of great importance to the ecosystem and also to human life due to their ecological functions, including seed dispersal, pollination and control of both urban and agricultural pests (BONANÇA; SILVA, 2013). These animals are considered to be excellent bioindicators, demonstrating the quality and characteristics of the environment, making the diagnosis of studies related to environmental impacts quick and accurate, as they are easy to observe and identify (FILHO; MEDEIROS, 2006).

According to the Brazilian Committee of Ornithological Records (CBRO), 1901 species of birds were identified for Brazil in 2014, both resident and visiting, thus demonstrating the country's great ornithological wealth (MMA, 2014). Although some bird species live very well in anthropised environments, urban sprawl affects these vertebrates unevenly because it causes habitat destruction, but these animals are also the main targets of illegal hunting and wildlife trafficking (WWF, 2010). Therefore, the most effective way of conserving many species is still the conservation of environments and natural resources, together with the implementation of Environmental Education (GANEM, 2011; OLIVEIRA, 2011).

ENVIRONMENTAL EDUCATION (EA)

Talking about the environment, sustainability and environmental education has become increasingly common. This is due to a real understanding of the importance of environmental education, as people are gradually realising that the current way of life is unsustainable. Due to the impacts caused by global warming and the extinction of species, this importance needs to be matured and communication strategies developed to raise awareness of environmental education (GANEM, 2011; OLIVEIRA, 2011).

According to Carvalho (2004) and Minimi (2000), the term "Environmental Education" cannot be understood simply as something that approaches good practices within the environment, or even ecologically appropriate behaviour, but rather as a process that consists of providing people with a critical understanding of the biosphere and the environment. Thus, according to Nascimento (2011), environmental education is considered an extraordinary tool for raising awareness of the need to conserve biodiversity.

Carvalho (2004) also points out that environmental education is the only instrument capable of awakening new thoughts and behaviour, since individuals are able to reflect on their place within the perceived landscape, making it possible to evaluate their behaviour and change their actions in relation to it. Based on these findings, it is believed that the school environment is one of the most privileged and important places for carrying out environmental education. In contemporary times, schools have become increasingly responsible for transmitting social values and behaviours such as solidarity, respect and care for the environment (GOLDBERG et al. 2005).

Considering environmental issues, the school must offer effective means for each student to understand the phenomena that occur in the natural environment, human actions, and these as consequences for themselves, as well as for other living organisms and the physical environment itself (NASCIMENTO, 2011). However, the process of Environmental Education at school requires innovative practices capable of broadening perception, promoting a critical and self-critical sense, rescuing values and producing change (PALMER, 2006).

Among the various ways that teachers can work with their students in their pedagogical practices, we highlight the observation method for this work, which can be used to work with any group of animals, and in various areas of science and biology teaching, be it zoology, ecology, behaviour, or even biodiversity in general and environmental education (VIEIRA-DA-ROCHA; MOLIN, 2008). Within this proposal, we emphasise that birds are the best group to work with in environmental education with children, adolescents and/or the general public, because these animals are present in many places and are easy to appreciate, at least when compared to other vertebrate groups. They also stand out for these activities because of their beauty, variety and abundance of species, and because of the interest they naturally arouse in people (OLIVEIRA; TORRES, 2008; BECKER; POVALUK, 2013).

ENVIRONMENTAL EDUCATION AND BIRDS

According to Mohr and Moser (2009), the reasons for working with birds in

Environmental Education are diverse, mentioning the ease of identifying the animals, the great wealth contained in Brazil, the importance of this group of animals and of course the great acceptance of the population. According to Alenspach and Zuim (2013), many projects have been developed in the area since 1990, showing that the public most involved are schoolchildren, but not restricted to these environments, as there are also projects involving tourists, the elderly, rural communities and people who keep birds in captivity.

One of the most important aims of using bird activities in environmental education is to help participants realise the existence of animals that are close to human beings, especially vertebrates (OLIVEIRA; TORRES, 2008). By using birds as triggers for this process, it is possible to minimise the population's repulsion that the presence of wild animals in cities is dangerous, harmful, disgusting and unnecessary (ARGEL-DE-OLIVEIRA, 1996).

Thus, from the moment that individuals begin to consider it normal, or even desirable, for the population to live with native urban birds, a new movement begins to combat intolerance and repulsion towards other animal groups that are unfairly maligned (bats and amphibians) and various groups of non-synanthropic insects (bees, butterflies, beetles, mosquitoes, etc.) (vieira-da-rocha; mola, 2008; becker; povaluk, 2013).) (VIEIRA-DA-ROCHA; MOLIN, 2008; BECKER; POVALUK, 2013).

METHODOLOGY

The study was carried out between April and November 2015 at the Professor José Fernandes Machado State School, located in the neighbourhood of Ponta Negra, Natal/RN. The school in question has a close relationship with the natural environment, as it is located on the edge of the Lagoinha Environmental Protection Zone (ZPA5). In certain areas of the school it is close to the ecosystems of fixed dunes and tableland vegetation.

The target audience for the activity were 2nd year high school students, as they were studying vertebrates in class and were already familiar with them from previous activities during their compulsory internship. The activities were divided into three parts: (1) Presentation of the project; (2) Lecture; (3) Practical Activity.

In the first stage, called Presentation of the project, the objectives of the project and the action were mentioned and a diagnostic questionnaire (Table 1) was applied to analyse the students' prior knowledge of the topic that would be covered in the second stage.

The second stage was a direct lecture, with the aim of introducing the bird group, as well as its biology, morphology, diversity and behaviour, highlighting its conservation, importance and observation as an alternative for environmental education.

Table 1: First diagnostic questionnaire

	Questionnaire
1	Do you think birds can help humans? ( )YES ( )NO
2	Have you ever been birdwatching in the countryside? ( )YES ( )NO
3	Do all birds fly? ( )YES ( )NO
4	Do all birds sing? ( )YES ( )NO
5	Are there nocturnal birds? ( )YES ( )NO
6	Name five birds.

The first took place in area 1 (Figure 1A), which was chosen because it was the smallest anthropised spot and it was therefore possible to observe and identify the animals present there. At the chosen location there was only a sandy football pitch, surrounded by low vegetation, identified as tableland forest, delimited by the school wall, which makes contact with the protection zone, making it possible to visualise the tall vegetation. Area 2 (Figure 1B) was the school car park, which consisted of a more anthropised but more wooded environment, thus favouring the greater occurrence of animals. At the end of the second day of observation, the students were given an evaluation questionnaire, making it possible to analyse their understanding of the content (Table 2).

Figure 1: (A) Area 1 - Sand field with low vegetation all around. (B) Area 2 - School car park.

Table 2: Second evaluation questionnaire

	Questionnaire
1	Why were birds chosen for environmental education? What is the importance of these animals?
2	Would you take part in a birdwatching project outside of school hours? Please explain.
3	**"Birds are everywhere!"**. Comment on *this* statement, taking into account the observations made at the school.

| 4 | How many species of birds did you spot? Which ones? |
| 5 | How would you reconcile the study of birds with other subjects? |

Data was collected using two methodologies: visual recording, which is considered a qualified and complementary method of census by direct observation (RODRIGUES et al., 2005); and auditory recording, a very effective method due to the fact that most species emit unique sounds (REIS, 2011). Visual recording is a conventional method in ornithological studies for qualitative recognition (STRAUBE & URBEN-FILHO, 2005). Records were made using binoculars and photographic equipment (NIKON D3200).

Vocalisations were recorded simply with a mobile phone. Visual identification was carried out with the help of specific literature, such as the field guides (NETO, 2001; BINI, 2014a; BINI, 2014b; BINI, 2014c; GWYNNE, 2010). At the end of the sampling, the information was compiled into tables and the birds observed were grouped according to their taxonomic nomenclature. The main purpose of the auditory recording was to document the species that were not visualised.

RESULTS AND DISCUSSIONS

Sixteen students took part in the activity, and the diagnostic questionnaire showed that most of them had a low level of knowledge about birds and their importance. Of all the participants, the majority replied that birds are important for the environment and for human beings, but during the second moment, the lecture, not everyone could say how. Of the participating students, only three had already been on birdwatching trips, showing greater interest and helping the other students, arousing their interest even more.

The vast majority demonstrated knowledge of the main characteristics of birds, but even so, when asked to name five birds, 18 species and one order (Passeriformes) were mentioned within the bird class, as well as two animals not belonging to the group (Figure 2).

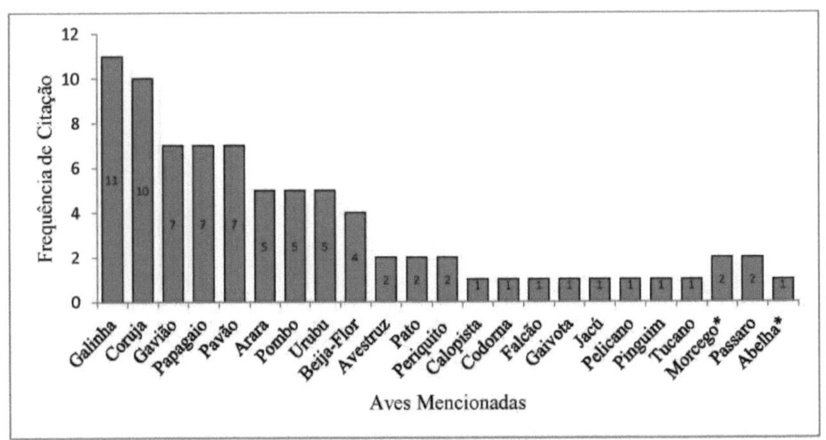

Figure 2: Birds mentioned by the students.

Still on the subject of species, it is possible to see that there is little knowledge of the great diversity of birds, as many students shared their knowledge when answering the question. As with Hanzem (2012), there is little knowledge, but great interest on the part of many students.

The second meeting, the lecture, was prepared according to the answers given in the diagnostic questionnaire. It explained the diversity of niches, the diversity of Brazilian birdlife, behaviours and the fact that flying is not an exclusive characteristic of birds, explaining why bats and insects are not representatives of the class.

During this time, it was possible to affirm the students' interest in birds by questioning their behaviour, mentioning a very common myth: "Is it true that every time someone is going to die, the owl warns"? According to Lopes and Santos (2004) popular knowledge can often coincide with the "reality of the species", but others are just myths.

The first observation, carried out in area 1, was the most productive in terms of the number of animals observed, due to the time of day, as it was carried out during the first class hour (7.00am - 8.40am). According to Bonança and Silva (2013), the time when birds are most active is between 6am and 10am. During this practical, seven species were observed, and each species was recorded (Table 3) and explained about its characteristics and habits, incorporating the knowledge into the practical activity and clearing up doubts from the questionnaire and the lecture, just like Lopes and Santos (2004). As well as observing the animals, it was also possible to see an old burrowing owl (Athene *cunicularia)*, which was found by the students themselves.

Table 1 - Record of birds observed in the practical activity

Family	Scientific Name	Popular Name
Cathartidae	*Cathartes aura* (Linnaeus, 1758) *Coragyps atratus* (Bechstein, 1793)	Red-headed vulture Black-headed vulture
Accipitridae	*Rupornis magnirostris* (Gmelin, 1788)	Red-tailed hawk
Columbidae	*Columbinapicui* (Temminck, 1813)	Piccolo
Cuculidae	*Crotophaga ani* Linnaeus, 1758 *Guira guira* (Gmelin, 1788)	Black anu White Anu
Estrildidae	*Estrilda astrid* (Linnaeus, 1758)	Lackey beak
Paseridae	*Passer domesticus* (Linnaeus, 1758)	Sparrow
Tyrannidae	*Myiozetetes similis* (Spix, 1825) *Pitangus sulphuratus* (Linaeus, 1766) *MaxOstorvie rixosa* (Vieillot, 1819)	Little red nose Birds of a feather Horse Suiriri

The second observation (Figure 3), which took place in area 2, was not as rich in quantity, but there was a great variety of birds, as it took place after the break (9.50am - 10.40am). The great variety observed not only aroused the students' interest even more, but also tested the knowledge they had absorbed in the previous practice, as they began to recognise the species they had already seen and even the songs of some of them. Despite the area being very busy, 11 species were recorded, 7 of which had already been observed in area 1. Nests were also observed at this time, but in trees, so it wasn't possible to identify which species they belonged to. Oliveira et. al (2015) mentions that the most urbanised areas have the least bird activity, even among synanthropic birds.

Figure 3: Students using binoculars (A) and a camera (B) to observe birds.

After the observations, the evaluation questionnaire was administered in the third stage, so that the students' learning could be analysed.

As for the importance of and the reason for choosing birds for the activity, the students referred to them as "excellent bioindicators", "animals sensitive to changes" and "possessors of unique niches", thus demonstrating that they really understood what they had done. They were also able to explain why there were fewer birds in the second observation.

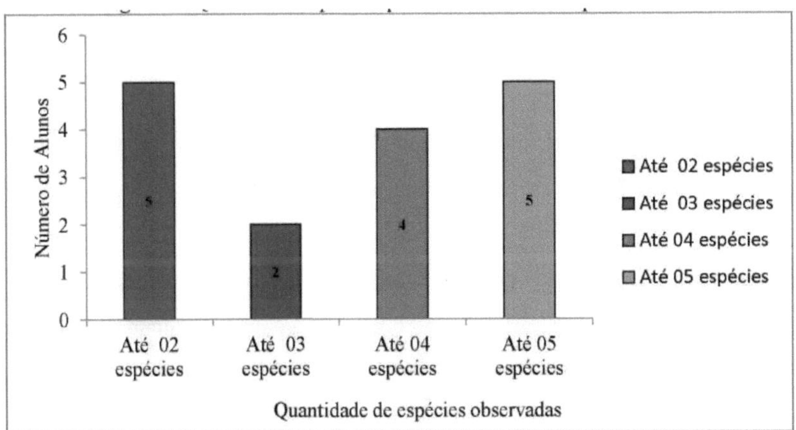

Figure 4: Number of species observed by the students.

Almost all of the participating students said they would do the activity again, except for one, who showed no interest and said he didn't have time. Even before they answered the questionnaire, they asked when the activity would take place again, and some suggested that it should take place outside of class and include other classes.

Only a few meetings were held, and just like Bastos (2006), the students were able to understand the importance of the topic, as the subject covered in the lecture was seen in practice, especially the relationship between the degradation of areas and the decline in species and specimens.

As with Hanzem (2012), by comparing the questionnaires before the lecture and after the observation activity, most of the class understood the content and used many of the terms they had learnt in class. It was also seen that the subject had been shared with other students. Vieira-da-Rocha & Molin (2008) also found that students said it would be possible to reconcile the subject of birds and birdwatching with various subjects, such as Science/Biology, Maths, Arts, History, Geography and Physics.

FINAL CONSIDERATIONS

This study has shown that birdwatching, as an innovative tool for teaching activities, is very effective as an alternative tool for environmental education, as it is an interdisciplinary teaching method. This is possible because practical activities arouse students' interest, thus facilitating the teaching-learning process and making knowledge solid. The level of interest, participation and satisfaction of the students was evident when we saw them talking about the species and having more and more questions.

The activity also made it possible to affirm that it can be carried out in any environment, from the most natural to the most urban, because the occurrence of birds is extensive, and it can not only be used for environmental education, but can also become a hobby.

With the results obtained, it is necessary to suggest this method, as well as better publicising it and possibly applying it in various schools, especially in communities close to Environmental Protection Zones and Conservation Areas, in order to sensitise them to the conservation of natural resources.

REFERENCES

ARGEL-DE-OLIVEIRA, M. M. Subsidies for the work of biologists in Environmental Education. The use of urban birds in environmental education. **World of Health**, V. 20, N°8, 1996. P. 263-270.

ALLENSPACH, N.; ZUIN, P. B. Aves como subsídio para a Educação Ambiental: perfil das iniciativas brasileiras. **Ornithological News Online**, No. 176, 2013.

BECKER, A. M.; POVALUK, M. Levantamento Das Espécies De Aves Da Área Denominada Zona de Preservação Ambiental e Lazer 1(Zpal1), Situada no Perímetro Urbano de Mafra - Sc. **Rev.Saúde Meio Ambiente**, v. 2, n. 1, p. 3-15, jan./jun. 2013.

BINI, E. **Birds of Brazil - Pantanal.** 1ª Edition, Homem-Passáros publications, Florianópolis, CS, 2014a.

. **Water birds of Brazil**. 1ª Edition, Homem-Passáros publications, Florianópolis, CS, 2014b.

______. **The colourful world of birds**. 1st Edition, Homem-Passáros publications, Florianópolis, CS, 2014c.

BONANÇA, R. A.; SILVA, A. M. Abundance and diversity of avifauna in four urban parks in the municipality of Jundiaí, SP analysed using geographic information systems. In: XVI Brazilian Symposium on Remote Sensing, **Proceedings**, Foz do Iguaçu, PR: INPE, 2013.

CARVALHO, I. **Educação ambiental e formação do sujeito ecológico.** São Paulo, Cortez, 2004.

BRAZILIAN COMMITTEE OF ORNITHOLOGICAL RECORDS - CBRO (2014). **Lists of the birds of Brazil.** 10th Edition. Available at <http://www.cbro.org.br>. Accessed on: 16 March 2015.

EFE, M. A. **Practical Guide to Birdwatching.** Santo Amaro da Imperatriz, 1999.

FILHO, J. A. de L.; MEDEIROS, M. A. S. Adverse impacts caused by urban afforestation activities. **Revista de Biologia e Ciências da Terra,** Universidade Federal de Sergipe, v. 6, n. 2, p. 375-390, July 2006.

GANEM, R. S. Conservation of Biodiversity: legislation and public policies. **Chamber of Deputies**, Chamber Editions, Brasilia, 2011.

GOLDBERG, R. A. T.; YUNES, M. A. M.; FREITAS, J. V. Children's drawing from the perspective of the ecology of human development. **Psicologia em estudo**, Maringá. V. 10, N.1; 2005. P. 97-16.

GWYNNE, J. A.; RIDGELY, R. S.; TYDOR, G.; ARGEL, M. **Aves do Brasil: pantanal & cerrado.** Editora Horizonte, São Paulo, 2010.

HANZEM, S. M.; GIMENES, M. R. **Importance of birds applied to environmental education in public schools in the municipality of ivinhema - ms**. Dourados, Uems, 2012. Available at : <http://anaisonline.uems.br/index.php/semex/article/view/582/587>. Accessed on: 19 June 2015.

LOPES, S. F.; SANTOS, R. J. **BIRDWATCHING: FROM ECOTOURISM TO ENVIRONMENTAL EDUCATION.** Caminhos de Geografia, Institute of Geography UFU. V. 5 N. 13; 2004. P. 103-121.

MINIMI, I. Teacher training in environmental education. Text on training in Environmental Education. **Panoramic Workshop on Environmental Education**, MECSEF - DPEF. Coordination of environmental education, Brasilia, 2000. P. 15-2.

MMA-Ministry of the Environment. **Red Book of Endangered Brazilian Fauna.** Brazil, 2014.

MOHR, M.; MOSER, G. **Birdwatching as a Tool for Environmental Education.** 2009.

NASCIMENTO, M. V. E. **Estudo Das Percepções Ambientais E De Ações Educativas Promotoras Da Biodiversidade Em Unidade de Conservação No Rio Grande Do Norte.** Natal, RN, 2011. Dissertation (Master's) - Federal University of Rio Grande do Norte. Biosciences Centre. Regional Postgraduate Programme in Development and the Environment/PRODEMA, 2011.

NETO, M. R. **Guia ilustrado:** Fauna da escola das Dunas de Pitangui - terrestrial ecosystems. Natal, Moura Ramos, 2001.

OLIVEIRA, D. V. The influence of urbanisation on the composition of the bird community in three areas of the city of Natal, RN. In: **62ª SBPC Annual Meeting.** Available at: < http://www.sbpcnet.org.br/livro/62ra/resumos/resumos/4459.htm>. Accessed on: 19 August 2015.

OLIVEIRA, E. S.; TORRES, D. F. **Environmental Education in the Genipabu Apa, How is it going?** Rev. eletrônica Mestr. Educ. Educ. ISSN 1517-1256, v. 21, 2008.

OLIVEIRA, E.; IRVING, M. de A. **Convention on Biological Diversity after Nagoya:** challenges for the media in a country of megadiversity. Razón y Palabra. N. 75, 2011.

PALMER, J. A. **50 Great Modern Educators:** From Piaget to Paulo Freire. São Paulo: Context, 2006. P. 326.

PEREIRA, K. D. L; SILVA, R. SURVEY OF THE AVIFAUNA OF THE URBAN AREA OF ANÁPOLIS, GOIÁS. **Ensaios e Ciência: Ciências Biológicas, Agrárias e da Saúde,** Anápolis, v. , n. 2, p.33-46, 2009. Available at: <http://www.redalyc.org/articulo.oa?id=26015684004>. Accessed on: 19 August 2015.

POUGH, H.; JANIS, C. M.; HEISER, J. B. **The Life of Vertebrates.** 6ª edition, Atheneu editora, 2003. P. 446.

REIS, R. S. L.; MACHADO, P. C. M. **Pre-processing of sound signals from nocturnal birds.** Goiânia, Goiás, 2011.

RODRIGUES, M.; CARRARA, L.A.; FARIA, L.P.; GOMES, H.B. **Aves do Parque Nacional da Serra do Cipó:** o Vale **do** Rio Cipó, Minas Gerais, Brazil. Rev. Bras. Zool., v.22, n.2, p. 326-338, 2005.

STRAUBE, F.C.; URBEN-FILHO, A. Avifauna of the Salto Morato Nature Reserve (Guaraqueçaba, Paraná). **Atualidades Ornitológicas On-line,** n.124, p.12, 2005.

VIEIRA-DA-ROCHA, M. C.; MOLIN, T. A aceitação da observação didática no ensino formal de aves como ferramenta didática no ensino formal. **Actualidades Ornitológicas On-line,** N° 146 - November/December 2008.

WWF - BRAZIL. **Bird Guide to the Atlantic Rainforest of São Paulo -** Serra do Mar and Serra de Paranapiacaba. 1ª edition, São Paulo, 2010.

CHAPTER 2

THE USE OF SOCIO-SCIENTIFIC ISSUES INVOLVING CONTROVERSIAL TOPICS FROM THE LIFE SCIENCES

Naama Pegado Ferreira[4]
Clécio Danilo Dias da Silva[5]

INTRODUCTION

The rapid advances in scientific production and technological artefacts have brought numerous controversies and controversial situations to the heart of society, especially with regard to the ethical, social, cultural, political, economic and special interest dimensions of the purposes to which these technologies are applied.

Among the countless controversies of today, which are presented to individuals on a daily basis, mainly through the media and information (media, internet, newspapers and magazines), are topics such as: the use of embryonic stem cells, cloning, abortion, the release of cannabis, among others, because these are complex issues that, because they involve aspects related to beliefs and values, often lead to a division of opinion among society, which gives them a controversial and controversial character.

Complementing this thought, Forgiarini and Auler (2009) state that socio-scientific issues (SSI) are controversial and are nothing more than issues with a considerable scientific/technological dimension, as well as commonly arising from the impacts that these innovations can have, generating disagreements between the academic community and society. These are usually issues that involve ethics and morality. This makes it somewhat challenging for teachers to address these issues in the classroom.

For this reason, there was a need to raise these discussions in biology classrooms in order to find out: What are the students' opinion(s) about different social and scientific issues in which they are involved? In addition, it is believed that socialising their positions and respecting other people's points of view is essential for an adequate and participatory civic education.

---

[4] Master in Teaching Natural Sciences and Mathematics (UFRN).
[5] Master in Teaching Natural Sciences and Mathematics (UFRN).

In this context, the aim of this work was to find out about the main socio-scientific themes present in the daily lives of 3rd year high school students at a public school in an urban area in Natal/RN, as well as to provide a moment of discussion for them to express their opinions on various contents, which is part of their autonomy of thought.

THEORETICAL FRAMEWORK

Some current studies show that socio-scientific issues are gradually being included in the school curriculum as a way of approaching controversial content in science classrooms, with the aim of favouring students' Scientific Literacy (CL) (REIS; GALVÃO, 2008). For these authors, there are some impediments to its realisation, such as: teachers' fear of controlling protest situations during discussions; their unpreparedness due to lack of knowledge in managing such discussions and also due to excessive content, assessments that condition the curriculum and do not consider these themes. Venau and Costa (2016) also reiterate that the professionals themselves believe that their colleagues are unprepared to deal with controversial issues in the classroom.

Even with comprehensive scientific production in the area of scientific literacy, there are still few and few works in this area of CSFs, especially in basic education, and when they are found, they usually have a diffuse and fragmented approach, as reported by Fernandes et al. (2015). This situation must be changed, as it is of great importance that CSFs are explored in the classroom, since they are part of students' daily lives and are directly related to their critical and civic education, so that there is truly meaningful learning. In addition, their use makes it possible to promote scientific literacy, as they co-operate in scientific conceptual understanding with the development of a more comprehensive view of science. It is also known that its use can make students more skilful in argumentation and reasoning, through participation in debates, helping them to become ethically and socially responsible.

Argumentation is also essential for students to consolidate their opinions and defend their points of view. In the literature, its importance is often cited in teaching and learning processes, especially in science teaching, although it is considered rare and somewhat complex (MCNAIL; PIMENTEL, 2010). There are a number of very complex issues in the biological field, which are therefore controversial because they involve: ethics, morals, beliefs, the values of each individual, religion, culture, the social and economic impacts they can have, and many others. These include: Is cloning ethical? What about caring for animals, should they be proliferated and only used in the service of science? Is the use of embryonic cells for study acceptable? Is abortion a woman's decision alone? Would the use of cannabis not only for therapeutic purposes reduce violence and drug trafficking? Are the realities different in our country?

In this context, Siqueira and Scheid (2015), when investigating the frequency of controversial themes in science and biology textbooks in Brazil, found that the most scarce themes basically refer to three dimensions: sustainability and emerging themes in science; environmental problems and the relationship between human beings and the environment. The authors also report that "in order to lay the foundations for this practice at school, it is essential that teachers from all areas of knowledge are well prepared, as they will be the ones who encourage and mediate these discussions" (SIQUEIRA; SCHEID, 2015, p.89). This corroborates the work of Razera and Nardi (2006), who point out the absence of discussions on ethics and moral development in research into science teaching, which is something that should be thought about and worked on in classrooms in order to reduce this gap in teaching.

METHODOLOGY

In order to carry out this investigation into the proposed topic, we opted to use the foundations of both quantitative and qualitative approaches, since according to Gressler (2007, p. 102): "qualitative research aims to understand a specific, ideographic reality, whose meanings are linked to a given context". The research was carried out at the end of the 2017 school year, with 31 students from the 3rd year of secondary school as the target audience, since they are usually at the end of basic education, about to take the National High School Exam (ENEM) and where they usually need to position themselves in relation to different contemporary issues. The intervention lasted 3 hours.

Initially, the students answered a questionnaire containing eight questions on current and controversial issues involving socio-scientific problems in the field of biology. It should be emphasised that none of these topics had previously been covered in class, so that the students could express their initial opinions without interference from others. The proposed questions were: *1) Are you in favour of or against abortion? 2) Are you in favour of or against embryonic stem cell studies? 3) Are you in favour of or against the consumption of transgenic foods? 4) Are you in favour of or against sex reassignment surgery via the Unified Health System (SUS)? Why? 5) Do you believe in life outside the Earth (extraterrestrial)? 6) Are you in favour of or against cloning? 7) Are you in favour of or against releasing the use of cannabis in Brazil? 8) Are you in favour or against racial quotas in Brazil?* Identification on the questionnaire was optional and justification for the choices was requested for all questions.

Afterwards, a conversation circle was held in which the students were able to socialise their opinions on the different subjects, where, with the teacher's mediation, they were able to expose their doubts and opinions to the other students. Everyone who wanted to speak signed up and gave their opinion in turn; those who didn't want to give their opinion were encouraged with a few questions; rhetoric on the part of the teacher, regardless of the student's position, was also used as a way of getting them to reason differently from their point of view, showing that there are always people who can think differently from us on the same subject. Finally, they were asked to research and choose one of the three topics below to produce an argumentative essay: Freeing ordinary citizens to carry guns in Brazil, legalising cannabis in Brazil or Legalising abortion in Brazil.

The above data was quantified and discussed according to the content analysis assumptions proposed by Bardin (1977). The themes were chosen by the teacher in advance as being relevant to learning biology and also socially conflicting and widely debated.

RESULTS AND DISCUSSIONS

1st moment (Questionnaire)

Of the 31 students who answered the questionnaire, only 13 wanted to identify themselves, as it was optional. We can see from this that many students still find it difficult to share their opinions, possibly due to the lack of activities in which they need to demonstrate their personal positions. According to Santos, Mortimer and Scott (2001), the need for teachers to develop pedagogical interventions that help increase students' argumentative capacity is essential when discussing CSFs. The data obtained through the questioning can be seen in figure 1. The students generally expressed their opinions on each topic using the following categories: "In favour", "Against", "In some cases", "I don't know".

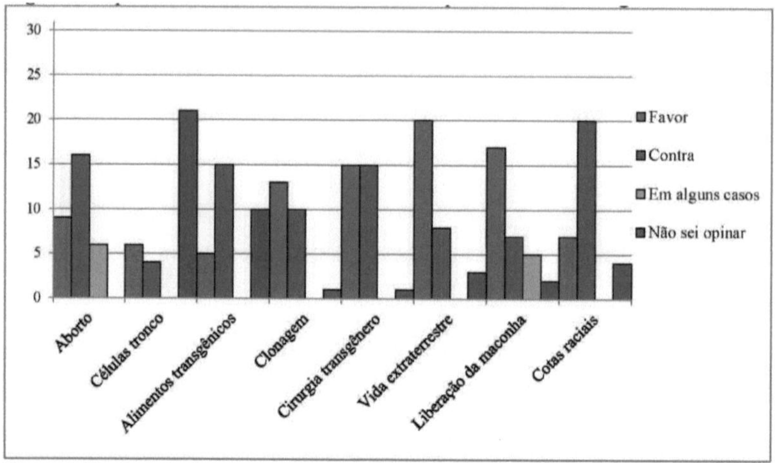

Figure 1: Students' opinions on controversial topics in Biology.

Embryonic stem cells was the topic with the most abstentions, as many of the students didn't know what it was about and preferred not to give an opinion. It is believed that this result is due to the lack of information on the subject during primary school, but also to the decline in controversy generated by the subject following the discovery of researcher Shinya Yamanaka, winner of the Nobel Prize in Physiology in 2012, for having succeeded in reprogramming adult cells to become totipotent.

The issue that most divided opinion was the transgender surgery offered by SUS. Only one student preferred not to give an opinion, the rest of the class were confused. This is possibly related to the frequency of debates, especially in the media, over the last few years. Some respected each other's sexual orientation, but didn't agree with the "use of *public resources for sex reassignment surgery"*. Other students argued that "there are *other, more serious problems to be dealt with in the SUS, because there are people who are at risk of death and need it much more than these people"*. In this context, Oliveira, Oliveira and Andrade (2016, p.128): "[...] among other recommendations, educational institutions should guarantee those who request the right to oral treatment exclusively by their social name", this is already a legal reality for transgender individuals, as well as the regulation of the surgical procedure by the Resolution of the Federal Council of Medicine. Although the debate on the subject is current and necessary, it is believed that it is still little discussed in classrooms, especially biology classrooms, because some teachers believe that individuals should only be studied for their genetic characteristics and differences between the biological sexes.

Another very controversial topic was abortion. Many students believe that "*abortion is illegal and that there are countless ways of preventing it", and* that the fact that[a] *an unwanted pregnancy can often be the result of irresponsibility on the part of the parents".* Some preferred not to give their opinion on rape because they said that "it's *hard to imagine yourself in the person's shoes and that a woman has the right to her own body".* There were also students who were against *the teachings of Jesus",* which shows a religious value in the formation of the individual that affects their choices and social behaviour.

With regard to legalising the use of cannabis, many students were totally in favour. It is believed that they have a lot of information about the harms and benefits that cannabis can bring, unlike other topics, possibly because it is part of their reality and daily lives.

Surprisingly, the students generally didn't know what GM food was and couldn't give their opinion. To fill this gap, these and other topics, such as embryonic stem cells, were explained before the discussions. According to Fonseca and Bobrowski (2015), with biotechnological advances, it is necessary to include new topics in the classroom and update the textbook, since it is the tool most used by teachers and the topics emphasised in these books are those that are commonly in the media.

Many students believe that the universe is too vast for life to exist on planet Earth alone, which is a question that has been studied extensively over the years in terms of technological advances and resources and space exploration. Some have reported having seen flying saucers and that they just need to be explored further to find them, while those who are against it use scientific arguments that there is no proof of this yet.

One of the most surprising themes in the discussions was that the majority of students were against racial quotas. Many said that "there are *black people who can afford it and people shouldn't be 'classified' by the colour of their skin and that only quotas related to family income should be released".* Others in favour said that "*it's the least Brazil can do for black people, given that it was the country with the longest history of slavery, more than 300 years of suffering".*

2nd moment (Discussion)

As they are controversial, it generated a lot of discussion on each point that was raised, but as the teacher has the role of mediator, he intended to discuss the previous topics without taking a position, letting the students show their opinions.

Some students initially found it difficult to accept other people's opinions, believing that only their arguments were valid. This was gradually deconstructed during the debate by listening to divergent opinions and the reasons why they thought differently. Expressing ideas is important for teenagers and also for learning; respecting other people's understanding of various issues, even if they can form their own opinion, and arguing is also a matter of education. Some studies suggest that spontaneous argument in science classes takes place in different ways, such as Munford and Teles (2015):

> When we investigated argumentation [...] using an alternative theoretical-methodological approach, we observed that in this classroom, students spontaneously argued in different ways when learning science. The high occurrence of unplanned argumentative situations can be attributed to the teacher's willingness to engage in dialogue and his more reflective professional attitude, which regulated his discursive interactions with the students (MUNFORD; TELES, 2015, p. 178).

Many students said that after the discussion they were able to realise other people's points of view and even change their minds. What is interesting to note is that some students specifically reported that they were based on their mum's ideas and concepts, meaning that even with all the autonomy of thought that school can provide, the family still plays a fundamental role in young people's education and decision-making.

3rd moment (textual production)

Of the 31 students, only 21 handed in the requested activity, of which 4 chose to address the issue of gun ownership, 4 the legalisation of cannabis and 12 the legalisation of abortion, as shown in the figure below.

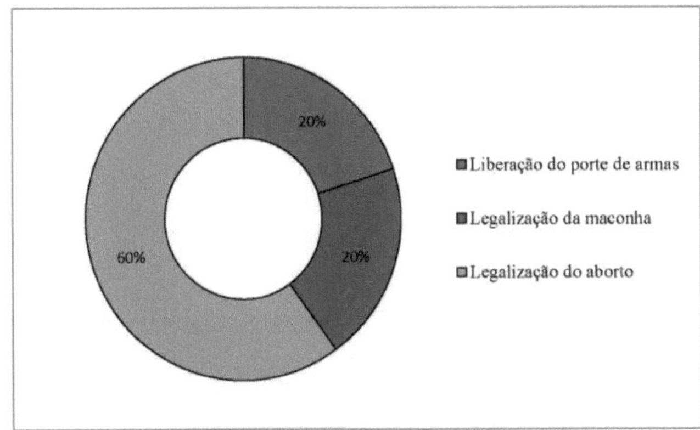

Figure 2: Students' choice of essay topics.

Of the 20% who chose to argue in favour of allowing people to carry guns, 100% were in favour, arguing that many people already "own *guns illegally"* (A.L.S, 16 years old), while others said that most people are afraid of robberies, exposed to risky situations at any time, and that carrying a *gun* would guarantee citizens "the *right to defend themselves and annul the differences in strength between the attacker and the victim"* (E.B.C.C., 17 years old). Santos (2016) points out that this issue is part of the daily life of the Brazilian population and is of paramount importance in the national legal scenario, even though it is not widely debated, it has recently come up for discussion again.

As for the topic of legalising cannabis, of the 20% who chose it, 100% were in favour of releasing and legalising cannabis. With a variety of arguments: from the use of taxes to improve health and education; savings in public security; a reduction in the trafficking of this drug, as in the USA; a reduction in the price of *cannabis-based* medicine in the country, since there would be no need to import it; as well as an increase in jobs, among others. One of the important points to highlight refers to a student's argument in which she mentions that: "the law doesn't *differentiate between users and dealers, and the police officer is responsible for that, and as we live in a society that is prejudiced to the extreme, the criteria for differentiating them, apart from the technical ones, is skin colour and social class"* (B.L.F.C., 16 years old) showing that the students perceive the social problem macroscopically and not just economically, associating it with much more serious problems such as poverty and racism.

Finally, the one that stood out the most when it came to choosing a topic for the text was abortion. It is believed that among the topics proposed, this is the one that generates the most controversy for young people, as it includes personal and moral values and is part of some people's reality, or with friends and close relatives who have already come across such a choice. Among the students who chose this topic, 69% were in favour of legalising abortion, 27% were against it and only 4% were in favour of releasing it in specific cases, such as malformation of the foetus and rape.

Those *in* favour said they found it more cruel *"the unbelievable number of children living on the streets, in orphanages and even without a father, which is nothing more than a male abortion,"* (N.C.G.S., 16) explaining another problem and social taboo which is male responsibility in this process of choice. According to Pirotta and Schor (2016), this topic generates more controversy than consensus and the meaning of the speeches is strongly marked by the process of social construction of gender roles. Other students argued that because we live in a democratic society we should have the right to free choice, even so, some of the students defended the implementation of public policies to prevent unwanted pregnancies and that it was also the responsibility of the state, which even if it legalised abortion did not *encourage it"* (E.F.N., 17 years old).

The students who were against this act argued that it *was* an *"inhumane* act *(...) no one has the right to take the life of another person, especially if that pregnancy was the result of two irresponsible people"* (M.M.B., 18) because it was something that could be prevented. They also mentioned *how a person can think they have the right to take a human life that is already being formed?"* (J.M.S.C., 17). Others are concerned beyond the factor of life, they see motherhood as something divine, *"it's having the gift, the grace, an inexplicable love that is becoming a mother (...) we are nobody to take the life of another human being"* (M.L.S.N., 18 years old). This shows that because of personal or even religious values, we wouldn't be able to choose or take responsibility for human life.

FINAL CONSIDERATIONS

In the light of the activities carried out, we realised that socio-scientific issues/themes are of great relevance to the formation of citizens, as well as contributing to the practice of respecting other people's positions, generating more appreciation and acceptance of others, on both sides, despite the fact that there are always different thoughts and disagreements in positions.

It can be seen that the students' level of argumentation and opinion-forming improved after the discussions, which can be consolidated through the exposition of their speeches and writing. We believe that there are other biological issues that deserve to be highlighted, such as: evolutionism, creationism, excessive consumerism, care for the environment, among others that arise on a daily basis, which can also be addressed together with the themes of this work, through new methodologies, providing a space for reflection and argumentation for the student, contributing to their autonomy, critical thinking and citizen training.

In future work, it is also proposed that students' levels of argumentation be categorised in order to better understand the evolution of their thinking. It is recommended that this be done using readings of articles and mock juries as a means of improving students' argumentation, as proposed in various works in the specialised literature.

REFERENCES

BARDIN, L. **Content Analysis**. Edition 70: Lisbon, 1977.

FERNANDES, L.L.; SILVA, E.M.; CORDEIRO, E.L.S.; PRATA, R.V. Working with Socio-scientific Issues in the Early Grades: a review of the literature in Science Teaching. In: 10 Encontro Nacional de Pesquisa em Ensino de Ciências, 2015, Águas de Lindoia. **Proceedings...** X ENPEC, Aguas de Lindoia, 2015.

FONSECA, V.B.; BOBROWSKI, V.L. Biotecnologia na escola: a inserção do tema nos livros didáticos de biologia. **Acta Scientiae**. v. 17, n.2 , p.496-509, May/Aug. 2015.

FORGIARINI, M.S.; AULER, D. The approach to controversial themes in Youth and Adult Education: the case of "afforestation" in Rio Grande do Sul. **Revista Electrónica de Ensenanza de las Ciencias.** v.8, n. 2, p. 399 - 421, 2009.

GRESSLER, L. A. **Introdução à pesquisa:** Projetos e relatórios / Gressler, L. A. - 3. ed. - São Paulo: Loyola, 2007. 328 p.

MCNEILL, K. L.; PIMENTEL, D. S. Scientific discourse in three urban classrooms: The role of the teacher in engaging high school students in argumentation. **Science Education**, Malden, v. 94, n. 2, p. 203-229, 2010.

MUNFORD, D.; TELES, A.P.S.S. Argumentation and the construction of learning opportunities in science classes. **Revista Ensaio**, Belo Horizonte, v. 17, n. Especial, p.161-185, nov. 2015.

OLIVEIRA, E.Q.O.; OLIVEIRA, P.B.Q.; ANDRADE, G. B.O. Transsexual surgery: medical, legal and social reality. **Revista Jurídica da Escola Superior do Ministério Público de São Paulo**, São Paulo, v. 10, n.2, p. 115-130, 2016.

PINZINO, W. **Socioscientific Issues: A Path Towards Advanced Scientific Literacy and Improved Conceptual Understanding of Socially Controversial**

Scientific Theories Dean.2012. 39 f. (Dissertation). Specialisation in Education. University of South Florida, Tampa, 2012.

RAZERA, J.C.C; NARDI, R. Ethics in science teaching: responsibilities and commitments to children's moral development in discussions of controversial issues. **Investigações em Ensino de Ciências,** Porto Alegre, v.11, n.1, p. 53-66. 2006.

REIS, P.; GALVÃO, C. Natural Science teachers and the discussion of social-scientific controversies: two distinct cases. **Revista Electrónica de Ensenanza de las Ciencias.** v.7, n.3, p. 746 - 772, 2008.

SANTOS, C.V.L. **O fracasso do estatuto do desarmamento.** 2015. 18 f. (Final Coursework). Bachelor of Laws. Tiradentes University, Aracaju, 2015.

SANTOS, L.P.W.; MORTIMER, E.F.; SCOT, P.H. Argumentation in socio-scientific discussions: reflections from a case study. **Revista Brasileira de Pesquisa em Ensino de Ciências**, Belo Horizonte, v.1, n.1, p. 14 -26. 2011.

PIROTTA, K.C.M.; SCHOR, N. Considerations on the voluntary interruption of pregnancy based on the discourse of USP university students. In: 14th National Meeting of Population Studies, 2016, Caxambú. **Proceedings...** XIVABEP, Caxambú, 2016.

SIQUEIRA, A.C.; SCHEID, N.M.J. A abordagem dos temas controversos em livros didáticos de ciências e de biologia brasileiros. **Revista Interacções,** Lisbon, v. 11, n. 39, p. 69-91. 2015.

VENEU, F.; COSTA, M. Controversial topics in the classroom: ^are we listed? A small investigation among science teachers. **Revista Electrónica Interuniversitaria de Formación del Profesorado**, v. 19 n.2, p.89-100. 2016.

CHAPTER 3

THE SCIENCE-TECHNOLOGY-SOCIETY APPROACH IN EDUCATIONAL RESEARCH

Clécio Danilo Dias da Silva[6]

INTRODUCTION

According to Auler and Bazzo (2001) and Teixeira (2011), the Science-Technology-Society (STS) movement emerged in the mid-20th century as a response to dissatisfaction with the traditional conception of science and technology, the political and economic problems related to scientific and technological development and environmental degradation. Since then, science teaching has increasingly been thought of in conjunction with issues that relate science, technology, society and the environment, and academic production on CTS is growing (MIRANDA, 2013).

The emergence of CTS studies has been recorded internationally since the 1980s and nationally since 1990, with the publication of the first Brazilian master's thesis defended in the field of CTS education in 1992, and from then on it has grown rapidly (CACHAPUZ et al., 2008). According to Miranda (2013), over the last 20 years, in the Brazilian context, the CTS field has been consolidating and this can be verified by the growing accumulation of productions in international and national journals, in the proceedings of academic events, and in theses and dissertations.

According to Freitas and Ghedin (2015) it is due to this rapid development of the CTS field that it is becoming increasingly essential to know the national academic production, implying a periodic review of such production, identifying its theoretical-methodological assumptions, trends, research objectives and themes, main results and possible contributions to improving teaching and training, as well as the development of new fields of research.

Based on this concern, the aim of this work was to carry out a "State of the Art"

---

6 Master in Teaching Natural Sciences and Mathematics (UFRN).

investigation into the Science, Technology and Society approach in the papers published in the editions of the National Education Congresses (CONEDU) between 2014 and 2016, with a view to identifying the main results, trends and possible contributions to improving science teaching and developing new lines of research.

METHODOLOGY

In order to map and evaluate the production of academic research on the CTS Approach in the editions of the National Education Congresses, we used the research modality characterised as "State of the Art", using the qualitative approach to understand the information found in this study, the nature of the productions presented, the general characteristics and the trends seen in the productions written on the subject under study. According to Luna (2011), state-of-the-art research seeks to describe the current state of a particular area of research and is an excellent source of updates for the field of research in the area and/or subject under study, as it condenses the most important topics of the problem of this area and/or subject under study and generally presents, in addition to what is already known, the main gaps and theoretical and/or methodological obstacles. Ferreira (2002) points out that this research makes it possible to recognise dominant and emerging themes and approaches, as well as gaps and unexplored fields available for future research.

As a methodology for analysis, we used the elements of Content Analysis systematised by Bardin (2011). By analysing the content of a text, we can gather quantitative or qualitative indicators about the production of the work. This method of analysis uses systematic and objective procedures to describe the content and its indicators. This analysis is carried out in a number of stages, in this case, the first stage is to choose which documents to analyse, the second stage is to categorise these documents and to conclude the Content Analysis, the interpretation of the quantitative and qualitative data consists of conclusions that are pertinent to the objectives of the research taking place.

Initially, we searched the pages of the event's proceedings (http://www.editorarealize.com.br/revistas/conedu/anaisanteriores.php) using the keyword **"CTS Approach" to find** all possible occurrences. The 03 (three) editions of **CONEDU'S**

(2014 - 2016) were fully investigated in the search for papers to analyse. To select the sample, the following criteria were followed: the title and/or keywords of the work must expressly contain the expression "Science-Technology-Society", "CTS" and their equivalents.

In order to structure the analysis, it was considered that a diversity of themes explored by the researchers was obtained, which made it possible to organise the following categorical groupings: Distribution by edition of the event, teaching modalities and subjects investigated, Thematic Focus (Implementation of the CTS approach, Conceptions of CTS, CTS in Textbooks, Literature Review), Methodologies and Techniques (Two or more techniques combined; Theoretical study/literature review; Written and oral interviews and questionnaires; Focus group, Observation, Unspecified).

The percentages of the biggest trends found in the categories analysed were calculated, and a basic descriptive statistical analysis was carried out on all the material collected. In this way, the distributions were identified, thus determining the probable trends of the categories found.

RESULTS AND DISCUSSION

The results obtained in this research show that of the 6,339 papers published in the annals of the National Education Congresses, only 17 papers presented the use of the Science, Technology and Society Approach, representing 0.80% of the event's publications (Table 1). This shows that although there is consolidation in the use of the STS approach, there is still a deficit in the number of academic papers when compared to the total amount of research in the educational field. Similar results were found by Abreu et al. (2009), who, when analysing publications in national journals, found that academic production using CTS was still not very significant (0.78%) in relation to the total production of published works, as in the present study.

Chart 1: Number of papers emphasising the CTS approach at CONEDU'S.

YEAR	EVENT	LOCAL	ARTICLES	ARTICLES WITH A CTS APPROACH	%
2014	I CONEDU	Campina Grande/PB	1421	02	0.14
2015	II CONEDU	Campina Grande/PB	2020	10	0.49
2016	III CONEDU	Natal/RN	2898	05	0.17
TOTAL			**6339**	**17**	**0.80**

With regard to the "Thematic Focus" of each study, 11 papers were found to focus on the Implementation of the CTS Approach (65%), 04 on Conceptions of CTS (23%), 01 on Literature Review (06%), and 01 was related to the analysis of Textbooks (06%) (Table 2). These results allow us to verify that the thematic focus "implementation of the CTS approach" in the classroom has been receiving more attention from the researchers participating in the CONEDU editions. This fact is intrinsically interesting, since it responds to a concern raised by researchers in the field regarding the implementation of the CTS proposal in the school context (FREITAS; GHEDIN, 2015).

This fact can be seen in the research carried out by Cachapuz et al. (2008) and Hunsche et al. (2009), who, when developing studies on the state of the art of the CTS Approach, identified in international (1993 to 2002) and national (1998 to 2008) research, respectively, the predominance of studies that discussed the theoretical assumptions of the CTS field and little research reflecting on effective implementations in science teaching, revealing the lack of studies on the intervention of the CTS approach in the school context.

Chart 2: Expressiveness of the thematic focus in research emphasising the CTS approach at **CONEDLTS.**

THEMATIC FOCUS	I CONEDU	II CONEDU	III CONEDU	TOTAL	%
Implementing the CTS Approach	02	04	05	11	65
Conceptions of CTS		04		04	23
Literature review/theorists		01		01	06
Textbooks		01		01	06
TOTAL				17	100

For the category "Training Modalities and Subjects Investigated," we found that 10 papers were aimed at secondary school students and teachers (59%) and 02 at primary school teachers (12%), making primary education (71%) the most evident modality in CONEDU publications. This was followed by four papers on higher education (23%) and one on all types of education (06%).

Chart 3: Methodologies and techniques used in works emphasising the CTS approach at **CONEDU'S.**

METHODOLOGIES AND TECHNIQUES	I CONEDU	II CONEDU	III CONEDU	TOTAL	%
Two or more techniques	01	01	02	06	35

Combined					
Theoretical study/literature review		03	01	04	23
Interviews or Questionnaires	01	03		04	23
Focus group		01	01	02	12
Observation		01		01	07
TOTAL				**17**	**100**

Similar results were found by Miranda (2013) in Brazilian theses and dissertations, in which 49% were aimed at primary education, with a predominance of research aimed at secondary education compared to primary education. According to the author, this predominance of research in secondary education may be related to the fact that most of the references that discuss CTS relations are in the areas of Physics, Chemistry and Biology, which are the subject areas belonging to secondary education.

With regard to "methodologies and techniques", we found that 06 papers used two or more techniques combined to construct data (35%), 04 carried out interviews or applied a questionnaire (23%), 04 developed theoretical studies/literature reviews (23%), followed by 02 papers involving focus groups (12%) and 01 with observation (07%). Analyses have allowed us to see that the methodologies and techniques follow the most significant thematic focuses (implementations, theoretical reviews and conceptions), which represents theoretical and methodological coherence, i.e. the research methods adopted in the studies are consistent with their objectives. For example, most of the studies that sought to analyse the implementation of the CTS approach in the classroom used different collection techniques such as: observation, field diaries, recordings, analysis of constructed materials, questionnaires, interviews and focus groups.

According to Andrade (2007), the use of more than one data collection instrument, usually the interview combined with another, reveals the concern of researchers to look at the object under investigation from different perspectives, which seems appropriate in the field of education where the issues are generally very complex. Corroborating this thought, André et al. (2010) state that the combination of different forms of data collection represents an advance in research, since such a variety of collection sources indicates a broader approach to the issues, which brings greater richness to the research.

FINAL CONSIDERATIONS

Although the field of research on Science, Technology and Society has expanded considerably in the educational sphere in recent decades, it was observed that this is not reflected in the papers published at the National Education Congresses, as evidenced by the small number of studies involving this theme.

Based on the study carried out, it is possible to state that some trends in CTS research remain the same, such as "Training and Subjects Investigated", with the predominant application being in Basic Education (specifically Secondary Education), as found and discussed in various studies in this field.

There is also a predominance of "Methodologies and Techniques", with the use of a combination of two or more strategies, interviews and questionnaires. One difference that was noted in the CONEDU'S papers was the tendency towards "Research Focus", with a greater concentration on interventions/implementations of the CTS approach in the school context, and a lower concentration on theoretical essays and research into students' and teachers' conceptions.

In general, the results of this research constitute a contribution to reflection, debate, consolidation and advancement of research in the field of STS in local, regional and national contexts.

REFERENCES

ANDRADE, R. R. M. Research on teacher training: a comparison between the 1990s and 2000. In: ANPED Annual Meeting, 1, 2007. **Proceedings...** Caxambu: ANPED, 2007.

ANDRÉ, M. Formação de professores: a constituição de um campo de estudos. **Ciência & Educação**, v. 33, n. 3, p. 174-181. 2010.

ABREU, T. B.; FERNANDES, J. P; MARTINS, I. A qualitative and quantitative analysis of scientific production on CTS (science, technology and society) in science teaching journals in Brazil. In: Encontro Nacional de Pesquisa Em Educação Em Ciências, 7, 2009. **Proceedings of ENPEC.** Florianópolis: VII ENPEC, 2009.

AULER, D; BAZZO, W. A. Reflexões para a implementação do movimento CTS no contexto educacional brasileiro. **Ciência & Educação**, v.7, n.1, p.1-13, 2001.

CACHAPUZ, A; PAIXÃO, F; BERNARDINO LOPES, J; GUERRA, C. Do Estado da In: Encontro Nacional de Pesquisa Em Educação Em Ciências, 7, 2009. **Proceedings of ENPEC.** Florianópolis: VII ENPEC, 2009.

CACHAPUZ, A; PAIXÃO, F; BERNARDINO LOPES, J; GUERRA, C. Do Estado da Arte da Pesquisa em Educação em Ciências: Linhas de Pesquisa e o Caso "Ciência- Tecnologia-Sociedade". **Alexandria Revista de Educação em Ciência e Tecnologia**, n.1, p. 27-49, 2008.

FERREIRA, N. S. A. Research known as "state of the art". Educação e **Sociedade, v.** 23, n.79, 257-272, 2002.

FREIRTAS, L. M.; GHEDIN, E. Research on the State of the Art in STS: Comparative Analysis with the Production in National Journals. **Alexandria Revista de Educação em Ciência e Tecnologia**, v.8, n.3, p.03-25, 2015.

HUNSCHE, S. et al. The CTS approach in the Brazilian context: characterisation according to science education journals. In: National Meeting of Research in Science Education, 7, 2009. **Proceedings of ENPEC.** Florianópolis: VII ENPEC, 2009.

LUNA, S. V. **Planejamento de pesquisa:** uma introdução (2nd ed.). São Paulo, SP: EDUC, 2011.

MIRANDA, E. M. Analysis of the main trends in the science, technology and society (STS) perspective in Brazilian theses and dissertations in the areas of education and science teaching. **Ensenanza de las Ciencias**, v. 2013, p. 2214-2218, 2013.

TEIXEIRA, P. M. M. Educação científica e movimento CTS no quadro das tendências pedagógicas no Brasil. **Revista Brasileira de Pesquisa em Educação em Ciências**, v.3, n.1, p.23-41, 2011.

CHAPTER 4

APPLICATION OF A DIDACTIC SEQUENCE EXPLORING ECOLOGY AND ENVIRONMENTAL EDUCATION THEMES

Clécio Danilo Dias da Silva[7]

INTRODUCTION

According to Motokane and Trivelato (1999), Ecology is a relatively new science that has peculiar definitions and themes, but is of great importance for positioning in the world, and consequently involves themes that teachers should know well in order to work better with their students. Verona and Lima (2016) emphasise that studying ecology makes students understand the connection between human beings and everything around them and, consequently, the need to preserve the resources we have.

As knowledge of ecology is directly linked to issues of ecosystem functioning, it is extremely important that children, adolescents and young people learn its basic principles and theoretical foundations at school so that this subject can be worked on correctly (SILVA, 2012; FERREIRA; DIAS-DA-SILVA, 2017).

Content related to this topic is taught in schools throughout primary school, mainly in the 6th and 7th years of primary school and in the 1st year of secondary school, as well as being seen as a cross-cutting environmental theme in the Ministry of Education and Culture's (MEC) proposal in the National Curriculum Parameters (BRASIL, 1999). In this context, Lignani and Azevedo (2013) state that in various curricular proposals it is possible to find as a learning objective to develop basic scientific knowledge and also values about the need to preserve.

Despite this, teaching ecology doesn't just deal with directly ecological issues; it's important to develop situations that enable students to become critical citizens who are aware of their rights and duties in relation to environmental issues. In this way, the teacher's great challenge is to enable students to develop the necessary skills to understand the role of man

---

[7] Master in Teaching Natural Sciences and Mathematics (UFRN).

in nature (DIAS-DA-SILVA, 2018; NASCIMENTO et al., 2018).

The guidelines in the PCNs demonstrate the need to reconstruct the relationship between man and nature in order to definitively overturn the belief that man is the master of nature and is alien to it (FERREIRA; DIAS-DA-SILVA, 2017). By proposing the study of ecology associated with environmental issues, the PCNs seek to broaden students' knowledge of the structures and dynamics of nature and the ways in which life is processed so that students understand and realise that they are part of nature, understand how it is structured and how it works. In this way, students will identify with their problems and be able to act more consciously (SILVA, 2012, VERONA, LIMA, 2016).

In this context, this work aimed to explore basic knowledge of ecology and environmental education through a didactic sequence in science teaching, contributing to the critical and reflective training of students in relation to the conservation of biodiversity and the environment.

METHODOLOGY

This research is qualitative in nature, using the Participatory Research (PP) approach, since it is inserted in the educational context and has the figure of the educator/researcher as a component who has been part of the scenario studied for three years, teaching the subject of Science and Biology. Demo (1999, p. 126) defines PP research as: "a research process in which the

The community participates in analysing its own reality, with a view to promoting social transformation for the benefit of the participants".

The research was carried out between February and March 2018, with three 6th grade primary school classes (a total of 107 students) from a public school located in the urban area of Natal, Rio Grande do Norte, Brazil. The work was based on the educational dynamics of the Three Pedagogical Moments according to the assumptions of Delizoicov, Angotti and Pernambuco (2002, 2011).

In the **First Pedagogical Moment** (initial problematisation), a problematising question was used to probe the students' knowledge. The question used was: During a science field lesson in *Ponta Uegra, Ualiana saw a crab trying to capture a small fish from the water. The* *girl quickly scared off the crab to protect the defenceless fish, ensuring its survival. Do you agree with the action taken by the student? Can you relate it to any of the themes explored in class?* The students wrote down their positions in their notebooks and then socialised them with the whole class, instigating discussions in the classroom.

During the **Second Pedagogical Moment** (organisation of knowledge), dialogic classes and conversation circles were held to explore the contents of Ecology (Fig. 1). In order to make knowledge more attractive and meaningful, the film "Bug's life" was shown. This is an animated film produced by *Pixar* in 1998 and distributed worldwide by *Walt Disney Pictures*. In the film's plot, every year a group of grasshoppers demand a share of the ants' harvest. But when something goes wrong and the crop is accidentally destroyed, the grasshoppers threaten to attack and the ants are forced to ask for help from other insects to face them in a "battle". Considering the diversity of ecological concepts dealt with in the animation, at the end of the film, discussions were held on the main scenes involving the theme. At another point, the text "Nature and human action" by Pena (2018) was read to encourage discussions about the impact of human beings on the environment.

In the **Third Pedagogical Moment** (application of knowledge), some activities were developed involving the selected content and the initial problematising questions. In order to provide moments of retention and fixation of the *"basic concepts of ecology", the* students were asked to build Concept Maps using the initial content discussed in class. To work on the theme*" ecological relationships"* the class was divided into 6 groups and they were instructed to make posters on the sub-themes selected by the educator/researcher (colony, society; commensalism; tenancy; pedantry; parasitism). The posters were then socialised with the class.

Afterwards, in order to work on the sub-themes of *"food* chain*", the* students were taken to an open space in the school to take part in a dynamic "food chain", which demonstrated the organisation and interlinking of a food chain and the importance of the representatives in the flow of energy and in the ecosystem. Initially, some boards were distributed, with one student representing a sun and the other participants symbolising an animal. Then the representative of the sun started the activity by throwing a roll of string to another representative of a certain photosynthesising animal. Following this dynamic, the roll was thrown to a herbivore and then to a carnivore. The latter would then throw it to a decomposer, which would return to the sun to restart the cycle. When the roll was passed on to the decomposer, the educator/researcher intervened to explain and clarify any doubts about the influence of that individual's death on the entire food chain.

RESULTS AND DISCUSSIONS

During the initial problematisation, we found that most of the students were touched by the action taken by the fictional character Juliana in "helping the defenceless fish against the crab". The students used various arguments, such as:

> "She *did the right thing, the fish was in trouble, from what I understood it was small and defenceless and was trapped"* (A. G. S. M., 11 years old);
>
> *"Juliana did the right thing, if I were in her shoes I would do the same, the crab could very well look for another way to feed itself, (Dava (hitting it because he realised it was small)"* (D. C. D. S., 10 years old);
>
> *"It was a kind thing to do and shows that she cares about nature"* (M. C. O., 11 years old).

This situation is worrying because it shows that the students were unable to relate to any of the science topics, such as the ecological relationships that are established between living beings and even how the food chain works. However, it should be emphasised that this moment of questioning and discussion enabled us to assess the students' prior knowledge and structure the other stages of the educational process.

Freire (1987) states that problematisation takes place through dialogue; the starting point for it is the critical and reflective analysis that conscious subjects carry out on a significant dimension of tangible reality, which is presented to them as a problem, and they can seek and construct answers. According to Gehlen (2009), it is through problematisation that students are challenged to expose their understandings of certain significant situations, concepts and understandings of the subject matter.

It is in this line of reasoning that Ferreira andias-da-Silva (2017) state that it is through the initial survey that the teacher is able to identify the prior knowledge that the students have, so that they can adapt their teaching planning according to the learning needs of the class, while also making it possible to monitor and evaluate the students' progress during their teaching activities.

During the dialogic classes and conversation circles in the organisation of knowledge, it was found that the students' interest in ecology content was piqued, intensifying the learning process. Activities involving dialogues and discussions promote students' cognitive development, as well as contributing to the organisation and, consequently, learning of science content, which helps students to deal with information, understand and re-elaborate it, and thus understand and interact with the world and act autonomously in it (DIAS-DA-SILVA et al., 2016; 2018; NASCIMENTO et al., 2018).

Also with regard to the organisation of knowledge, we highlight the use of the film "insect life", as this resource enabled students to better understand ecological relationships and the functioning of the food chain within natural ecosystems. As a result, we can assure you that films/animations can be used in the classroom as a way of interacting with the content programmed in the teaching work plan, especially in the areas of Science and Biology, which have a diversity of concepts to be assimilated. In this context, some authors have validated the literature and indicate the use of the film "insect life" to explore various elements in the content of taxonomy, zoology and ecology (FRANCO; SANTANA-REIS; LOPES, 2013; SANTOS; GEBARA, 2015; OLIVEIRA et al., 2016).

Through the application of knowledge, the students were able to put into practice the themes explored during the organisation of knowledge, in response to the prior knowledge identified in the initial questioning. The development of concept maps played an important role in this process, enabling the students to grasp the basic concepts of ecology and preparing them for the other activities (Fig. 2).

According to Ontoria Pena et al. (2005), concept maps enable meaningful learning to the extent that they are drawn up by students and used as a tool for appropriating knowledge. Completing this thought, Moreira (2010) states that the use of CMs allows students to delve deeper into the content covered, enabling them to determine the relationship between concepts, as well as to differentiate between more inclusive and less inclusive concepts.

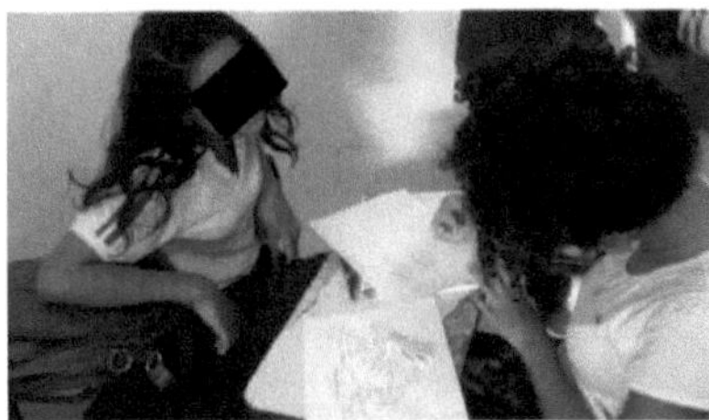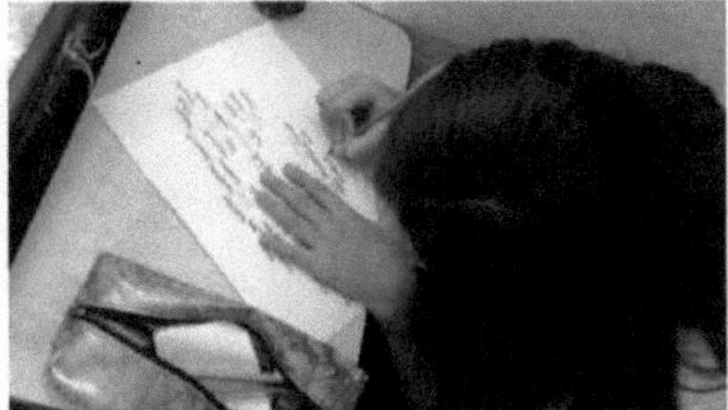
Figure 2: Drawing up concept maps involving the basic concepts of ecology.

The activity of making posters and, above all, socialising these materials with the class, enabled students to identify their conceptual errors and alternative conceptions about ecological relationships, favouring their collaborative learning (Fig. 3).

Figure 3: Making and socialising posters on Ecological Relationships.

According to Veiga (2000), teaching is socialised when it is centred on the student's intellectual action on the object of learning through cooperation between working groups, the teacher's directiveness, not only with the aim of facilitating learning, but also to make teaching more critical (making contradictions explicit) and creative (elaborate expression).

In this sense, both the teacher and the student cease to be passive subjects and become active subjects, capable of proposing coherent actions to overcome the difficulties detected (DIAS-DA-SILVA, 2018). It allows for the exchange of knowledge, stimulating the development of respect for ideas, critical thinking, questioning and solutions, favouring the exchange of experience, information, cooperation and mutual respect between students, enabling meaningful learning.

During the dynamics of organising and linking ideas, the students participated actively, demonstrating knowledge and mastery of the content taught in the round table discussions. This was noticeable during the course of the discussion, as the students knew the components that structure food chains and defined their relationships within ecosystems.

According to Campos (2010), dynamics are methodological tools that favour students' ability to form concepts, relate ideas, establish logical relationships and develop oral expression. For the author, they also enable initiative, reflective capacity, teamwork skills, assimilation, problem-solving and the development of autonomy on the part of students.

In general, we agree with Verona and Lima (2016) when they say that studying ecology is not just about understanding conceptual aspects. Learning about this area of knowledge should be explored in conjunction with elements of Environmental Education, since human beings only take care of what they know. When we don't study something, we can't value its conservation. It is in this context that we believe that the application of a didactic sequence on ecology, exploring topics related to the environment, contributed to the students' critical and reflective training in environmental issues.

FINAL CONSIDERATIONS

Although ecology is widely taught in the classroom, it is still common for it to be taught in a traditional and mechanistic way, which often leads to superficial learning, based on the simple memorisation of concepts. This situation is aggravated by the fact that these contents are not explored in a contextualised way in relation to the students' real lives. In order to change this scenario, this research, in line with the recommendations of the PCN's (BRASIL, 1996, 1998), explored the contents of ecology associated with the environment, involving everything from basic concepts and ecological relationships between living beings to man's action on nature.

The positive results found through a teaching sequence involving problematising questions, conversation circles, the application/discussion of films, the construction of concept maps, dynamics and the creation and socialisation of posters, highlight its potential for working on ecology content in science classes, contributing to the critical and reflective training of students in the conservation of natural resources.

REFERENCES

BRAZIL. Ministry of Education and Sports. **The "National Curriculum Parameters" and primary education.** Brasília: MEC/SEF, 1996.

BRAZIL. Ministry of Education and Sport. **National Curriculum Parameters**: presentation of cross-cutting themes. Brasília: MEC/SEF, 1998.

CAMPOS, L. M. L. Group dynamics as didactic-pedagogical tools for secondary biology teaching. In: VIII EDUCERE. 4. 2010. Curitiba, Paraná. **Proceedings...** PUCPR-Paraná, 2010.

DELIZOICOV, D.; ANGOTTI, J.; PERNAMBUCO, M. M. **Ensino de Ciências**: fundamentos e métodos. São Paulo: Cortez, 2002.

DELIZOICOV, D.; ANGOTTI, J.; PERNAMBUCO, M. M. **Ensino de Ciências**: fundamentos e métodos. 4. ed. São Paulo: Cortez, 2011.

DEMO, P. Methodological elements of participant research. In: BRANDÃO, C. R. **Repensando a pesquisa participante**. São Paulo: Brasiliense, 1999.

DIAS-DA-SILVA, C. D. et al. Approaching the respiratory system from a perspective of the three pedagogical moments. **CARPE DIEM: Cultural and Scientific Journal of UNIFACEX**, v. 16, n. 1, p. 29-43, 2018.

DIAS-DA-SILVA, C. D. et al. Learning about the human body: contributions of pibid to science teaching. **CARPE DIEM: Cultural and Scientific Journal of UNIFACEX**, v. 14, n. 1, p. 17-30, 2016.

DIAS-DA-SILVA, C. D. Potentialities of the Agroecological Garden in the Path of Environmental and Food Education. **Unisanta BioScience**, v. 7, n. 1, p. 1-5, 2018.

FERREIRA, N. P.; DIAS-DA-SILVA, C. D. **Educational practices in the teaching of Science and Biology**: didactic proposals for basic education. Germany: New Academic Editions, 2017.

FRANCO, I.R.; SANTANA-REIS, V.P.G.; LOPES, P.P. Entomology in cinema: analysis of the film vida de inseto as a didactic resource. In: Brazilian Symposium on Cultural Entomology, 1, 2013. **Proceedings...** Feira de Santana: I SBEC, 2013.

FREIRE. P. **Pedagogy of the oppressed.** 17. ed. Rio de Janeiro: Paz e Terra, 1987.

GEHLEN, S. T. **The role of problems in the science teaching-learning process: contributions from Freire and Vygotsky**. 2009. 253 f. Thesis (Doctorate in Scientific and Technological Education) - Federal University of Santa Catarina, Florianópolis, 2009.

GEHLEN, S. T; MALDANER, O. A; DELIZOICOV, D. Pedagogical moments and the stages of the study situation: complementarities and contributions to science education. **Ciência & Educação**, v. 18, n. 1, p. 1-22, 2012.

LIGNANI, L. de B.; AZEVEDO, M.J. C. Whose "Home"? Environmental History and the Teaching of Ecology. In: Encontro Nacional De Pesquisa Em Educação Em Ciências, 9., 2013, Águas de Lindóia. **Proceedings of ENPEC...** Águas de Lindóia: ABRAPEC, 2013.

MOREIRA, M. A. Concept maps as tools to promote progressive conceptual differentiation and integrative reconciliation. **Ciência e Cultura**, v.32, n.4, p.474-479, 2010.

MOTOKANE, M.T.; TRIVELATO, S.L.F. Reflexões sobre o Ensino de Ecologia no Ensino Médio. In: Encontro Nacional De Pesquisa Em Educação em Ciências, 2., 1999, Valinhos. **Proceedings of ENPEC...** Valinhos: ABRAPEC, 1999.

NASCIMENTO, A. C. L. M. et al. Practical activities in science teaching: the relationship between theory and practice and the training of undergraduate students in biological sciences. **CARPE DIEM: Cultural and Scientific Journal of UNIFACEX**, v. 16, n. 1, p. 44-60, 2018.

OLIVEIRA, A.B.R. et al. Analysis of the animated film "Vida de Inseto" in the light of Animal Biology. In: Rio de Janeiro Entomology Symposium, 3, 2016. **Proceedings...** Rio de Janeiro: UNIRIO, 2016.

ONTORIA PENA, A. et al. **Concept Maps**: a technique for learning. São Paulo: Loyola, 2005.

PENA, R. F. A. "**Nature and human action**"; Brasil Escola, 2018. Available at: <u>https</u>://brasilescola.uol.com.br/geografLa/natureza-acao-humana.htm>. Accessed on 26 April 2018.

SANTOS, J. N.; GEBARA, M. J. F. Pedagogical Analysis of Films: Genre of Animation in Science Teaching. **Colloquium Humanarum**, v. 12, n. 2, p.34-41, 2015.

SILVA, M. C. **Teaching ecology**: difficulties encountered and a working proposal for primary and secondary school teachers in João Pessoa, PB. 2012; 63 f. Monograph (Degree in Biological Sciences), Federal University of Paraíba, João Pessoa, PB, 2012.

VEIGA, I. P. A. **Teaching techniques**: Why not? Campinas: Papirus. 2000.

VERONA, M. F.; LIMA, E. S. A brief overview of ecology teaching based on the papers presented at the 5th national biology teaching meeting. **Revista da Sociedade Brasileira de Ensino de Biologia**, v. 6, n. 9, p. 3814-3824, 2016.

CHAPTER 5

A STUDY OF PRIMARY SCHOOL STUDENTS' KNOWLEDGE OF BACTERIA

Clécio Danilo Dias da Silva[8]

INTRODUCTION

According to Pozo and Crespo (2009), alternative conceptions are understood as individual subjective constructions created to explain natural phenomena, originating from individuals' everyday interactions with the world around them (CRESPO, 2009). According to Cândido et al. (2015), these conceptions are constructed by students from birth and accompany them into the classroom, where scientific concepts are inserted into the individual's cognitive structure. Corroborating this thought, Teixeira (2012) states that many sources of alternative conceptions are speculative at best.

The student's worldview is strongly influenced by their social environment, directly contributing to the formulation of concepts based solely on common sense. Most of the time, these derive from the interpretation of new experiences in the light of previous ones, and new concepts are impregnated with previous notions (PAIVA; MARTINS, 2017). Students' prior knowledge actively interacts with that studied formally at school, creating a range of unintended learning outcomes. Individuals' different conceptions of aspects related to health and disease prevention are closely linked to the culture in which students are inserted, through which they interpret the world. This intensifies the influence of everyday life on the scientific knowledge assimilated by the students' cognitive structure (BRUM, 2014; BRUM; SILVA, 2015).

With regard to microbiology, there has been a large number of studies aimed at assessing students' conceptions of bacteria (BEZERRA et al., 2009; SILVEIRA; OLIVEIROS; ARAÚJO, 2011; ARAÚJO; LOBATO, 2013; ARAÚJO; MEDEIROS, 2014;

---

8 Master in Teaching Natural Sciences and Mathematics (UFRN).

AZEVEDO; SODRÉ- NETO, 2014; CANDIDO et al, 2015; TOLEDO et al., 2015; SOUZA et al., 2016; OLIVEIRA, AZEVEDO; SODRÉ-NETO, 2016), however, few studies have been carried out to identify students' conceptions of bacteria and their implications for human health (ZOMPERO, 2009; BRUM, 2014; BRUM; SILVA, 2015).

With this in mind, the aim of this study was to analyse the alternative conceptions of primary schools students about bacteria, with the aim of contributing to the proposal of new pedagogical strategies for teaching science in basic education.

METHODOLOGY

The research was carried out with 30 7th grade students from an educational institution located in the North Zone of Natal, Rio Grande do Norte, Brazil. Considering the nature of the data collected, qualitative research was used using Bardin's (2011) content analysis procedures. According to the author, content analysis consists of a set of techniques for analysing communications in order to obtain, through systematic and objective procedures for describing the content of messages, indicators, quantitative or not, that allow the inference of knowledge.

In order to identify possible alternative conceptions, the students were asked to draw a picture according to their "understanding" of bacteria. The time allotted for drawing was free. It is worth emphasising that no prior explanations of the subjects were given so as not to influence what the participants actually had as mental representations of bacteria.

The drawings drawn by the students were analysed and categorised into groups to identify their level of understanding of bacteria. The classification used was adapted from Araújo and Medeiros (2014), Table 1.

Chart 1: Characterisation of the categories established from the analysis of the students' drawings.

Categories	Category characteristics
Category A: No design	The participants left the question blank.
CategoryB : Written Representations	The participants wrote something justifying their reasons for not drawing.
Category C: Concepts Alternatives and mistakes Concepts	The drawings didn't correspond to the bacteria and/or showed some kind of misconception about them.
CategoryD : Partial Representations	The drawings presented some conceptually acceptable representation of bacteria, such as their shape, structures

and emphasis on microscopic size.

Source: Adapted from Araújo and Medeiros (2014).

The data obtained was entered and grouped into tables in the Microsoft Excel 2010 application. The percentages of the major trends found in the categories analysed were calculated and a descriptive statistical analysis was made of all the material collected. In this way, the distributions were identified, thus determining the probable trends of the categories found.

RESULTS AND DISCUSSION

The students' conceptions of bacteria obtained from the drawings can be seen in Figure 1. It was observed that the majority of the drawings were included in Category D (55 per cent), which correspond to partially correct representations, since the drawings showed acceptable elements about bacteria, such as shapes (cocci, bacilli, spirochetes, flagellated cells, etc.), structures (cell wall, flagella, ribosomes, etc.) and size (Figure 2).

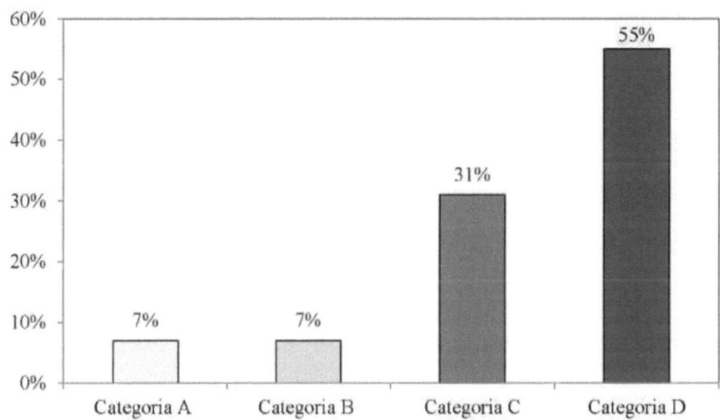

Figure 1: General analysis of the students' drawings about bacteria.
Key: Category A (No Drawing), Category B (Written Representations), Category C (Alternative Conceptions and Conceptual Errors), Category D (Partial Representations).

Seven (7%) of the drawings were included in Category A, as the students did not draw any pictures, and 7% were included in Category B, where they gave reasons for not doing what had been asked. According to Bahar et al. (2008), despite the advantages of using drawings to find out if there are alternative conceptions, the personal limitations of each student's drawing skills can have a significant influence on the results, as they may fail to draw either because they don't know how to express this knowledge or because they feel discouraged from trying. This can be seen in the students' written representations/justifications for not drawing, which were:

> AI *AND* *"I think I've seen some, but I can't remember what bacteria look like and I don't know how to draw them".*
> AI 7: *"I can't draw a bacterium because I've never seen one and because it can't be seen with the naked eye."*

However, the use of drawings and images can reflect students' mental models and cognitive representations, and can bring together characteristics, properties and impressions of an object or event that is understood, imagined or an abstract concept (BUCKLEY; BOULTER; GILBERT, 1997; MOREIRA, 2002).

Thirty-one (31 per cent) of the drawings were placed in Category C because many of the representations did not correspond to bacteria and/or showed some kind of conceptual misunderstanding about them.

Within this category, there was a large number of anthropomorphic representations, in which the bacteria had eyes, a mouth and/or a nose. It is believed that the attribution of human characteristics and behaviours to the various organisms (plants, animals, fungi, bacteria, etc.) is possibly related to the media (television, internet and electronic games) that are part of students' daily lives. In an attempt to make the content more attractive, they use personalisation full of exaggerations and attributes that are not consistent with the biological aspects of the various organisms present in the biosphere.

The lack of scientific knowledge about microorganisms makes it possible for the media, popular imagination and other sources of information - other than formal science education - to be the main sources of knowledge about these organisms for students in basic education, which justifies the presence of conceptual errors and alternative conceptions in the drawings produced. Similar results were found by Medeiros (2012), Araújo and Lobato (2013), Araújo and Medeiros (2014), Oliveira, Azevedo and Sodré-Neto (2016), when they analysed students' conceptions of microorganisms.

In basic education, content relating to knowledge of bacteria and their relationship with health is included in the subject of natural sciences and biology. Research has shown that this knowledge is often approached in a way that is decontextualised from the students' daily lives, and without any connection to health, environmental and technological issues, which has hindered learning (CASSANTI et al., 2008; KIMURA et al., 2013). Many studies point to the need to develop strategies or situations that encourage contextualisation, understanding and meaningful learning about microorganisms, enabling students to perceive their presence in their daily lives and their relationship with health and the environment (TOLEDO et al., 2015; SILVA, BASTOS, 2012).

FINAL CONSIDERATIONS

The results obtained in this research show that, although most of the students presented knowledge that was considered to be in line with what is scientifically acceptable regarding the various aspects of Bacteria (shapes, structures and size), many drawings showed conceptual errors and alternative conceptions about these organisms.

In this context, the work is a point of reflection for teaching practice, especially for microbiology, a subject that has scientific concepts and terminology that can hinder the learning process. The survey of previous ideas should be used as an initial tool in the teaching and learning process, as it provides the conditions for the teacher to identify what the student already knows and to work on the basis of these concepts.

REFERENCES

ALBUQUERQUE, G. G.; BRAGA, R. P. S.; GOMES, V. Students' knowledge about microorganisms and their use in everyday life. **Revista de Educação, Ciências e Matemática,** v. 2, n. 1, p. 12-25, 2012.

ARAÚJO, M. F. F.; LOBATO, L. S. Perceptions of protozoa in primary education: a diagnosis in schools in a coastal region of the Brazilian Northeast. **Acta Scientiae**, v. 15, n. 2, p. 354-362, 2013.

ARAÚJO, M. F. F.; MEDEIROS, M. L. Q. Alternative conceptions of teachers and students of basic education about protozoa, revealed by drawings, in schools of a semi-arid region of northeastern Brazil. **Revista da Sociedade Brasileira de Ensino de Biologia**, v. 5, n. 1, p.

5227-5238, 2014.

AUSUBEL, D. P. **Acquisition and Retention of Knowledge**: a cognitive perspective. Lisbon: Plátano, 2003.

AZEVEDO, T. M.; SODRÉ-NETO, L. Basic education students' knowledge about bacteria: scientific knowledge and alternative conceptions. **Revista de Educação, Ciências e Matemática,** v. 4, n. 2, p. 12-26, 2014.

BAHAR, M. et al. Science Student Teachers' Ideas of the heart. **Journal of Baltic Science Education**, v. 7, n. 2, p. 78-85, 2008.

BARDIN, L. **Content analysis**. São Paulo: Editions 70, 2011.

BIZERRA, A. et al. Young children and their knowledge of microorganisms. In: Encontro Nacional Em Pesquisa Em Educação Em Ciências, 7., 2009, Florianópolis. Proceedings of ENPEC... Belo Horizonte: ABRAPEC, 2009.

BRAZIL. Secretariat of Basic Education. **National Curriculum Parameters:** third and fourth cycles: presentation of transversal themes. Brasília: MECSEF, 1998.

BRUM, W. P. The Bacteria Theme in Primary Education: students' alternative conceptions of the implications for human health. **Revista de Ensino de Ciências e Engenharia,** v. 5, n. 2, p. 29-44, 2014.

BRUM, W. P; SILVA, S. C. R. Elementary school students' conceptions of bacteria and their relationship with human health. **Revista Ciências & Ideias,** v. 6, n. 2, p. 60-70, 2015.

BUCKLEY, B., BOULTER, C. GILBERT, J. Towards a typology of models for science education. In J. Gilbert (Ed.), **Exploring Models and Modelling in Science and Technology Education**, 90-105, Reading: University of Reading New Bulmershe Papers. 1997.

CANDEIAS, J. M. G. et al. **The use of educational games in the teaching of microbiology in primary and secondary schools**. Botucatu: UNESP, 2005.

CÂNDIDO, M. S. C. et al. Microbiology in High School: analysing the reality and suggesting teaching alternatives in a school in Paraíba. **Ensino, Saúde e Ambiente,** v. 8, n. 1, p. 57-73, 2015.

CARLÉTTI, D. S. Students' conceptions of microbiology. **Revista Brasileira da Sociedade de Ensino de Biologia,** v. 2, n. 1, p. 5112-5123, 2007.

CARVALHO, J. C. Q.; COUTO, S. G; BOSSOLAN, N. R. S. Some conceptions of high school students about proteins. **Ciência & Educação**, v. 18, n. 4, p. 897-912, 2012.

CASSANTI, A. C. et al. **Democratic**: teaching-learning strategies and teacher training. São Paulo: Colégio Dante Alighieri, 2008.

CASTRO, D. R.; BEJARANO, N. R. R. Prior knowledge about living beings of students in the initial grades of the Central Teaching Cooperative - COOPEC- BA. Revista Brasileira de Ensino de Ciências e Tecnologia, Ponta Grossa, v. 6, n. 1, p. 19-40, 2013.

FERREIRA, N. P.; DIAS-DA-SILVA, C. D. **Praticas educativas no ensino de Ciências e Biologia.** Germany: New Academic Editions, 2017.

GIL, A. C. **Como elaborar projetos de pesquisa**. Atlas, V. 4, São Paulo, 2007.

KIMURA, A. H. et al. Microbiology for secondary and technical education: the contribution of extension to the teaching and application of science. **Conexão UEPG Magazine**, v. 9, n. 2, p. 1326, 2013.

MARCONI, M. A.; LAKATOS, E. M. **Scientific Methodology**: science and scientific knowledge; scientific methods; theory, hypotheses and variables; legal methodology. São Paulo: Atlas, 2009.

MEDEIROS, M. L. Q. **Free-living protozoa in aquatic environments in RN: occurrence, characterisation and importance for basic education**. 2012. 75 f. Dissertation (Master's in Development and Environment), Federal University of Rio Grande do Norte, Natal, RN, 2012.

MOREIRA, M. A. Modelos mentales y modelos conceptuales en la ensenanza & aprendizaje de las ciencias. **Revista Brasileira de Pesquisa em Educação em Ciências**, v. 2, n. 3, p. 37-57, 2002.

OLIVEIRA, F. F.; AZEVEDO, T, M.; SODRÉ-NETO, L. Alternative conceptions about microorganisms: alert to the need for improvement in the biology teaching-learning process. **Revista de Ensino de Ciência e tecnologia**, v. 9, n. 1, p. 260-276, 2016.

PAIVA, A. L. B.; MARTINS, C. M. C. **Preconceptions of third-year high school students regarding topics in the area of Genetics**.Available:<http://www.fae.ufmg.br/ensaio/vol7especial/artigopaivaemarti ns.pdf>Accessed: 20.08.2017.

PELCKZAR, M. Microbiology. Vol 1, 2ª ed. Makron Books, 1996.

PIVATTO, W. Bacteria, viruses or fever: students' prior knowledge about health and combating diseases. **Revista de Educação, Ciências e Matemática,** v. 4, n. 3, p. 1-12, 2014.

POZO, J. I. **Aprendizaje de la ciencia y pensamiento causal**. Madrid: Visor, 1987.

POZO, J. I.; CRESPO, M. A. G. **Learning and teaching science**: from everyday knowledge to scientific knowledge." Porto Alegre: Artmed, 2009.

SILVEIRA, M. L.; OLIVEIROS, P. B.; ARAÚJO, M. F. F. Spontaneous conceptions of bacteria held by 6th to 9th grade students. National Meeting of Research in Science Teaching, 8, 2011. Proceedings **of ENPEC**. São Paulo, Campinas: ABRAPEC, 2011.

SIMONNEAUX, L. A study of pupils conceptions and reasoning in connection with micrpbes, as a contribution to research in biotechnology education. **International Journal of Science Education**, v. 22, n. 6, p. 19-32, 2000.

SILVA, M. S.; BASTOS, S. N. D. Teaching microbiology: perception of teachers and students in public schools in Mosqueiro, Belém, Pará. In: Encontro Nacional De Ensino De Ciências Da Saúde E Do Ambiente, 3., 2012 Niterói. **Proceedings...** Niterói: UFF. 2012.

SODRÉ-NETO, L.; DINIZ, J. A. Action research on teaching and learning microbiology in high school. **Ensino, Saúde e Ambiente,** v. 9, n. 2, p. 12-26, 2016.

SOUZA, F. F. et al. Alternative conceptions of high school students about bacteria. In: National Congress of Research and Teaching in Science, 1, 2016. Proceedings of **CONAPESC**. Paraíba, Campina Grande: Realise Eventos e Editora, 2016.

TEIXEIRA, A. M. M. B. **Alternative conceptions in science: a diagnostic tool.** 2011. 113 f. Dissertation (Master's in Biology Teaching), Faculty of Science and Technology, Lisbon, 2012.

TOLEDO, A. G. et al. Study of microbiology and its relationship to the student's daily life based on the theme of health. **Ensino, Saúde e Ambiente,** v. 8, n. 2, p. 23-42, 2015.

ZOMPERO, A. F. Elementary school students' conceptions of microorganisms in aspects involving health: implications for teaching and learning. **Experiências em Ensino de Ciências**, v. 4, n. 3, p. 31-42, 2009.

CHAPTER 6

EVALUATION OF A DIDACTIC UNIT INVOLVING THE PLATYHELMINTHES TAXON

Clécio Danilo Dias da Silva[9]

INTRODUCTION

Zoology is an area of science that studies animal diversity and basic knowledge is produced and organised through taxonomy, systematics and phylogeny. De Carvalho et al. (2007) emphasise that although the area of taxonomic knowledge is currently making progress, there is still a need for greater investment, specifically in the preparation of future specialists.

Teaching institutions at the most different academic levels, within the content related to biology, foster understanding of biological diversity, but it is up to the teacher to provide pedagogical guidance for exploring the taxa under study.

According to Marinho et al. (2012), there are many challenges to studying biological diversity, especially when it comes to developing learning about invertebrate taxa in the classroom. In this context, Pereira (2012) pointed out some problems that interfere with the quality of the teaching and learning process in this area, such as: poor initial training for teachers, which does not provide an adequate basis for working on the subject; a lack of practical lessons on zoology subjects, appropriate laboratories, a lack of teaching materials and a lack of knowledge of zoology teaching techniques.In view of this, Dias-da-Silva et al. (2017) state that the use of playful activities such as role-plays, games, modelling, schematic drawings, dynamic readings, the use of parodies, educational software and the creation of concept maps can help to minimise these problems, providing meaningful learning for zoological content.

According to Novak (2010), Concept Maps (CM) are graphic tools that enable the organisation of knowledge and represent significant relationships in the form of prepositions.

As visualised in Moreira (2010), concept maps are developed to show meaningful relationships between concepts taught in a single lesson, a unit of study or an entire course.

In this context, the aim of this work was to analyse a didactic unit involving the use of Concept Maps in the teaching and learning process of knowledge related to the Platyhelminthes taxon during basic education.

METHODOLOGY

[o]This research was carried out between July and August 2017 with 35 students from the 7th and 10th year of primary school at the "Master Colégio e Curso" educational institution, located in Rua Escritor Gilberto Amado, Pajuçara, North Zone of Natal - RN.

In order to make the teaching-learning process more meaningful, a didactic unit consisting of four stages was used. Firstly, there were lectures and dialogues with the aid of multimedia resources, enabling the deepening and systematisation of knowledge about the Platyhelminthes taxon, covering points relating to the morphology, physiology, ecology and phylogeny of these groups. At the end of each lesson, the content explored was summarised using previously prepared Concept Maps, with the aim of familiarising the students with the didactic resource, in line with the assumptions of Dias-da-Silva et al. (2017b). In the second stage, a process of familiarisation with concept maps was carried out as proposed by Trindade (2011), using readings and discussions of teaching material containing standards, steps and suggestions for constructing concept maps, prepared by Moreira (2010).

In the third stage, the students were asked to form groups (03 to 05 components) and draw up a concept map exploring the concepts studied in class about the taxon in question. In the fourth stage, the concept maps drawn up by the students were socialised by presenting and discussing them with the class. This moment enabled shared learning, allowing the students to identify the mistakes and successes of the material produced, favouring corrections of conceptual errors.

At the end of all the stages, the students were asked to evaluate the methodology used in the research and its contributions to the process of learning the content covered. To do this, we used a Likert scale (Table 1) as a collection tool. This material presented 05 statements about the activities carried out and a list of sentences for which the research subjects could express their degree of agreement by ticking: Agree (C); Disagree (NC); Indifferent (IN). According to Zanella, Seidiel and Lopes (2010), this resource is commonly used in surveys of attitudes, opinions and evaluations, and has made a significant contribution, adding reliability to research in the field of education.

The data obtained was grouped and categorised in tables in the Microsoft Excel 2010 application, in order to draw up graphs and tables and construct the results and discussions.

RESULTS AND DISCUSSION

Based on the answers obtained through the statements in the evaluation tool, it was possible to verify the students' satisfaction with the teaching sequence used (Chart 1).

For the first affirmative, 35 students (100%) agreed that "It is *easier and more interesting to learn zoology content using activities that make the lesson more dynamic"*. According to Dias-da-Silva et al. (2017), the use of alternative teaching methodologies attracts attention and piques students' interest

in understanding the various concepts dealt with in zoology, enabling more dynamic and meaningful learning of this content in the classroom.

Chart 2 - Evaluation of the teaching sequence involving the use of Concept Maps on the Platyhelminthes taxon. **Legend:** C (Agree), NC (Don't Agree), and IN (Indifferent).

AFFIRMATIVE	CT	NC	IN
1. It's easier and more interesting to learn zoology content using activities that make the lesson more dynamic.	35 (100%)	00 (0%)	00 (0%)
2. The lectures and dialogues helped me to understand the main characteristics of the Platyhelminthes taxon, such as morphological, physiological, ecological and evolutionary aspects.	26 (74%)	07 (20%)	02 (6%)
3. Discussions of didactic texts containing steps, norms and suggestions on how to draw up CMs helped me to understand and draw up my Concept Maps.	33 (94%)	00 (0%)	02 (06%)
4. Drawing up the concept maps enabled me to deepen my knowledge of the taxon covered.	34 (97%)	01 (03%)	00 (06%)
5. Group discussions about the MCs were important for building knowledge.	35 (100%)	00 (0%)	00 (0%)

With regard to the second statement, 26 students (74%) agreed that "The lectures *and dialogues helped me to understand the main characteristics of the Platyhelminthes taxon, such as morphological, physiological, ecological and evolutionary aspects."* 07 students (20%) did not agree, and 02 students (6%) were indifferent to the statement. According to Emerich (2010), activities involving dialogues, discussions and conversation circles facilitate students' cognitive development, as well as contributing to the learning of science and biology content, enabling the construction of scientific concepts aimed at developing skills that help students to deal with information, understand it, rework it, refute it, and thus understand the world and act autonomously in it.

In the third affirmative, 33 students (94%) agreed that *"Discussions of didactic texts containing steps, norms and suggestions on how to prepare CMs helped me to understand and prepare my Concept Maps",* and 02 students were indifferent (6%). According to Lourenço (2008), the process of familiarising students with concept maps allows them to get to know the basic structure of CMs, enabling them to better prepare this material. According to Trindade (2011), this stage is also relevant because it allows students to perceive CMs as a resource that contributes to learning the knowledge covered in class.

With regard to *the* fourth statement, 34 students (97%) agreed that "Drawing up the concept maps enabled me to deepen my knowledge of *the taxon covered"* and only 1 student (03%) disagreed with the statement. According to Ontoria Pena et al. (2005), concept maps enable meaningful learning to the extent that they are drawn up by the students and used as a tool for appropriating knowledge. In this sense, Lemos and Mendonça (2012), Primitivo et al. (2017) and Dias-da-Silva et al. (2017a) developed activities using concept maps to address zoological taxa, and found positive results, as was the case in this research.

For *THE* fifth statement, 30 students (100%) agreed that *"Group discussions about the CMs were important for building knowledge".* According to Moreira (2011, p.127) "Concept Maps must be explained by the person who makes them; by explaining them, the person externalises meanings. Therein lies the greatest value of a Concept Map".

According to Veiga (2000), teaching is socialised when it is centred on the student's intellectual action on the object of learning through cooperation between working groups, the teacher's directness, not only with the aim of facilitating learning, but also to make teaching more critical (making contradictions explicit) and creative (elaborate expression). It allows for the exchange of knowledge, stimulating the development of respect for ideas, critical reasoning, questioning and solutions, favouring the exchange of experience, information, cooperation and mutual respect between students, enabling meaningful learning (VEIGA, 2000).

FINAL CONSIDERATIONS

As a result of the positive aspects achieved in this research, the effectiveness of developing Concept Maps to explore biological content related to the Platyhelminthes taxon was verified. In this context, it can be inferred that CMs are a didactic resource that enhances meaningful learning about the taxa explored in Zoology. Based on this investigation, the use of this teaching tool becomes viable for working with different types of knowledge in the teaching of Science and Biology in basic education and allows us to add meaning from a student-centred teaching perspective.

REFERENCES

ARAÚJO-DE-ALMEIDA, E. (org.) **Ensino de Zoologia**: ensaios didáticos. João Pessoa: Editora Universitária, 2007.

ARAÚJO-DE-ALMEIDA, E. **Teaching Zoology**: metadisciplinary essays. João Pessoa: Editora Universitária, 2011.

AUSUBEL, D. P. **Acquisition and Retention of Knowledge:** a cognitive perspective. Lisbon: Plátano, 2003.

DE CARVALHO et al. Taxonomic impediment or impediment to taxonomy? A commentary on systematic and cybertaxonomic-automation paradigm. **Evolutionary Biology**, v. 34, n. 3-4, p. 140-143, 2007.

DIAS-DA-SILVA, C. D. et al. Use of concept maps in the teaching of invertebrate animals: contributions to the teaching of zoology in basic education. In: National Congress of Research and Teaching in Sciences, 2, 2017. Proceedings of **CONAPESC**. Campina Grande: Editora Realize, 2017a.

DIAS-DA-SILVA, C. D. et al. Concept maps and the learning of invertebrate taxa. In: FERREIRA, N. P.; DIAS-DA-SILVA, C. D. **Práticas educativas no ensino de Ciências e Biologia**. Germany: Novas Edições Acadêmicas, 2017b.

DIAS-DA-SILVA, C. D. et al. Zoological caravan: contributions to science and biology teaching. In: National Congress of Education, 3, 2016. Proceedings **of CONEDU**, Natal: Editora Realize, 2016.

EMERICH, C. M. **Science teaching**: a proposal to adapt knowledge to everyday life - a focus on water. 2010. 156 f. Dissertation (Master's in Education). Federal University of Rio Grande do Sul, 2010.

GOMES, A. P. et al. Teaching Science: Dialoguing with David Ausubel. **Revista Ciências & Ideias**, v.1, n.1, p. 23-31, 2010.

LEMOS, E. S.; MENDONÇA, C. A. S. Learning with concept maps: analysis of a didactic experience on the topic "Reptiles" with high school students. **Significant Learning in Review**, v.2, n.1, p.21-34, 2012.

LOURENÇO, A. B. **Analysis of concept maps drawn up by students in the 8th grade[a] of primary school based on lessons based on the theory of meaningful learning**: clay as a subject of study. 2008. 115 f. Dissertation (Master's Degree), UFSCar, São Carlos - 2008.

MARINHO, P. H. D. et al. Construction of a playful and innovative approach to learning the taxon Syndermata: the potential of a telejournalistic simulation. In: National Biology Teaching Meeting, 4, 2012. **ENEBIO Annals**. Goiânia, Goiás: SBEnBio, 2012.

MOREIRA, M. A. **Meaningful learning**: theory and complementary texts. São Paulo: Editora Livraria da Física, 2011.

MOREIRA, M. A. Concept maps as tools to promote progressive conceptual differentiation and integrative reconciliation. **Ciência e Cultura**, v.32, n.4, p.474-479, 2010.

NOVAK, J. D.; CANAS, A. J. The theory behind concept maps and how to make and use them. **Práxis Educativa**, v.5, n.1, p.9-29, 2010.

ONTORIA PENA, A. et al. **Concept Maps**: a technique for learning. São Paulo: Loyola, 2005.

PRIMITIVO, M. G. A. Conceptual map and playfulness in learning about the Nemertea taxon. In: National Congress on Research and Teaching in Science, 2, 2017. **Proceedings of CONAPESC.** Campina Grande: Realise Eventos e Editora, 2017.

VEIGA, I. P. A. **Teaching techniques**: Why not? Campinas: Papirus. 2000.

ZANELLA, A.; SEIDEL, E. J.; LOPES, L. F. D. Validation of a satisfaction questionnaire using factor analysis. **Revista de Inovação, Gestão e Produção,** v.2, n. 12, p.102-112, 2010.

CHAPTER 7

EXPERIMENTATION IN SCIENCE TEACHING: ESTABLISHING TRENDS AND DIALOGUES THROUGH SCIENTIFIC PRODUCTIONS

Clécio Danilo Dias da Silva[10]

INTRODUCTION

It's not new that students have a certain disdain for the natural sciences and that they have a distorted view, leading them to believe that there is no direct relationship between science (biology, physics and chemistry) and their daily lives. It's therefore up to the teacher to try to remedy the view that students have, as well as to improve their actions in the classroom, and experimentation is an alternative for streamlining lessons and boosting learning.

Experimentation in science teaching plays an extremely important role in the students' learning process, since subject areas such as biology, and especially physics and chemistry, have a great potential for mental abstraction in order to be understood. According to Guimarães (2009, p. 198) "in science teaching, experimentation can be an efficient strategy for creating real problems that allow for contextualisation and the stimulation of research questions".

The author also emphasises that "when teaching science at school, it must also be taken into account that every observation is not made in a conceptual vacuum, but on the basis of a body of theory that guides the observation". In this way, linking theory to practice can favour the process of building and developing knowledge. According to Carvalho, Batista and Ribeiro (2007): "Theory and practice should seek to merge into a unified and fruitful activity that enables students to permanently construct knowledge, and not just a brief memorisation of mathematical formulas and chemical equations for school assessment purposes (CARVALHO, RIBEIRO, 2007, p. 45).

---

10 Master in Teaching Natural Sciences and Maths (UFRN).

The idea that theory and experimentation should go hand in hand in the teaching and learning process can be applied in various ways, such as the classic way, where the student only observes the phenomenon, develops a method using a material and is not encouraged to discuss it. Another type of approach is one where experimentation is used to prove a concept, for example, the law is not questioned and the practice is used analogously to what was done to verify it. In addition to this, experimentation can play an integrating role with computer science and go hand in hand for the translation and analysis of different phenomena, as well as the production of the nature of scientific knowledge, i.e. the phenomenon is worked on before the concept and the student is encouraged to develop models to explain the observed facts and later those accepted by the scientific community are presented (SÉRÉ, COELHO; NUNES, 2003).

Therefore, it is up to the teacher to choose the possibility of applying experimentation in teaching. However, the choice would directly influence the role of the student during that lesson and that as teachers we should look for strategies that develop the cognitive actions of our students, favouring meaningful learning.

This paper provides an overview of the productions presented at previous editions of the National Education Congress (CONEDU) on experimentation in science teaching (biology, physics and chemistry), with the aim of analysing the occurrence and expressiveness of the work, highlighting the types of research, the level of teaching and the predominance of science subject areas.

METHODOLOGY

To develop the article, we opted to use qualitative research with Bardin's (2011) "content analysis" procedures, involving a systematic approach to data processing and analysis. According to the author, content analysis consists of "a set of techniques for analysing communications in order to obtain, through systematic and objective procedures for describing the content of messages, indicators (quantitative or not) that allow the inference of knowledge" (BARDEM, 2011, p. 44).

Bardin (2011, p.125) organises content analysis into three stages: I) Pre-analysis -

which is the actual organisation phase, corresponds to a period of intuition, but aims to make initial ideas operational and systematise them so as to lead to a precise outline of the development of successive operations in an analysis plan; II) Exploration of the material: consists essentially of coding, decomposition or enumeration operations, according to previously formulated rules, aggregating them into categories; and III) Treatment of the results: inference and interpretation, which consists of treating the raw results in such a way that they are meaningful ("speaking") and valid.

Initially, we searched the pages of the event's proceedings (http://www.editorarealize.com.br/revistas/conedu/anaisanteriores.php) for all possible occurrences of the theme Experimentation in Science Teaching. The 3 (three) editions of CODENU (2014 - 2015) were fully investigated in the search for papers to analyse. The following criteria were used to select the articles: they had to expressly contain the expression "Experimentation" in the title and/or keywords of the work, together with the terms "Learning", "Science Teaching", "Chemistry Teaching", "Physics Teaching" or "Biology Teaching". Subsequently, the selected materials were carefully read and explored, encouraging categorisation, inferences, treatment and interpretation of the data produced.

To define the categories of analysis, we used the assumptions of Bardin (2011, p. 147), who characterises them as "[...] an operation of classifying the constituent elements of a set by differentiation and then by regrouping according to genre (analogy), with previously defined criteria". Considering that there is a diversity of topics covered by the researchers, the papers were grouped as follows: I) According to the research approach
(Experience Reports, Teacher Training, Conceptions, Attitudes and Beliefs and Theoretical Essays and/or Reviews); II) Area of knowledge (Chemistry, Physics and Biology) and III) Level of target audience (Basic Education, Technical and Undergraduate).
In general, the data was grouped in Microsoft Office 2010 spreadsheets to create graphs, charts and tables for the construction of the results and discussions.

RESULTS AND DISCUSSION

Of the total of 6,339 papers published during the three editions of the National Education Congress, only 25 articles were published on the theme of "Experimentation in Science Teaching", representing 1.12% of all the papers published in the event's proceedings.

The number of papers published per event can be seen in Table 1.

Chart 1: Overview of the **CONEDU'S** annals (2014-2016), highlighting publications with an emphasis on Experimentation in Science Teaching.

YEAR	EVENT	LOCAL	ARTICLES	ARTICLES WITH EXPERIMENTATION	%
2014	I CONEDU	Campina Grande/PB	1421	02	0.14
2015	II CONEDU	Campina Grande/PB	2020	08	0.39
2016	III CONEDU	Natal/RN	2898	15	0.59
TOTAL			6339	25	1,12

With regard to the types of research involving experimentation in science teaching, we found that 15 articles were experience reports (60%), 04 articles were related to analysing conceptions, attitudes or beliefs (16%), 03 were aimed at initial or continuing teacher training (12%), and 03 works were theoretical essays or literature reviews (12%) (Figure 1). According to Suárez (2008), there have been many publications reporting positive experiences of teaching strategies applied in the classroom. According to the author, these reports, when contextualised with bibliographic references reflecting the teaching methodological direction adopted, become an important product for academic dissemination. For Rausch (2012), by reflecting on his practice, the teacher rethinks his actions and becomes a reflective teacher-researcher. Thus, it can be seen that teachers use their experience to produce research.

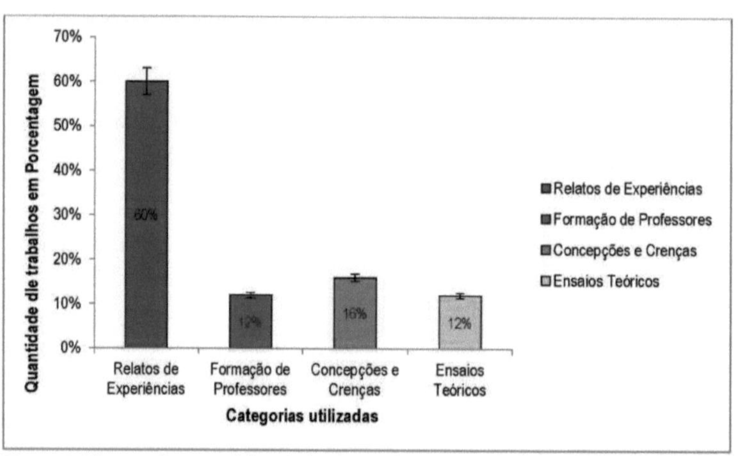

Figure 1: Types of research involving Experimentation in CONEDU'S works (2014-2016).

With regard to the areas in which the work was developed, 14 articles were carried out in the area of Chemistry (56%), 08 were applied in the area of Biology (32%) and 03 in the area of Physics (12%) (Figure 2). According to Silva (2016), in Brazil experimentation has taken on a prominent role in the various areas of the Natural Sciences, especially within the content explored in Chemistry. According to the author, this fact can be evidenced by the numerous publications addressing the theme of "experimentation in chemistry teaching" in event proceedings, journals, theses, dissertations, among others.

Due to the high degree of abstraction required to understand chemical content, experimentation becomes an ally that can favour the observation and association of chemical phenomena. According to Lisboa (2015, p.8) "Experimentation is one of the main foundations that support the complex conceptual network that structures chemistry teaching".

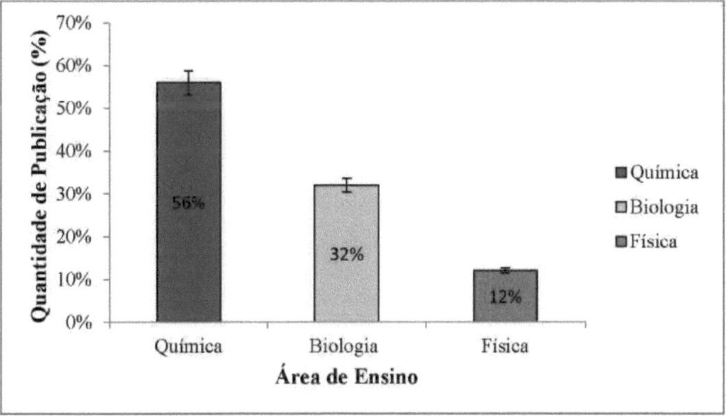

Figure 2: Teaching areas of papers involving the theme of Experimentation at CONEDU'S (2014-2016).

With regard to level of education, we found that 20 of the published articles were aimed at and/or involved primary education (52%), 03 articles were aimed at higher education (39%), and 02 at technical education (9%).

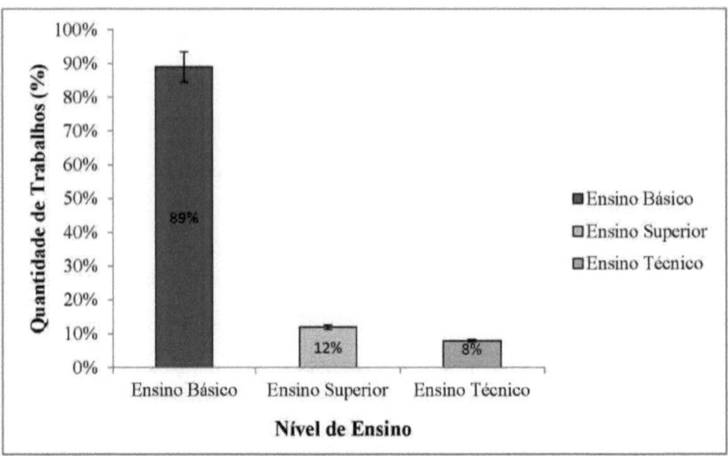

Figura 3: Expressiveness of the levels of education of the papers involving the theme of Experimentation at CONEDU'S (2014-2016).

According to Mayer (2013), considering that experimentation has numerous benefits in the learning environment, making students more motivated, it has been widely used by teachers in their pedagogical actions, gaining prominence at various levels of education, whether in basic education (primary and secondary), technical education or higher education. In this context, Oliveira and Trindade (2013) draw attention to the large concentration of research aimed at basic education involving practical lessons and experimental activities. According to the authors, the use of experimentation during basic education, whether in primary or secondary school, becomes favourable due to students' curiosity, giving teachers the opportunity to use this characteristic to carry out activities that provide learning, culminating in materials for academic dissemination.

FINAL CONSIDERATIONS

Despite the widespread use of experimental activities in pedagogical practices in the teaching-learning process, there are very few papers on the subject of "Experimentation in Science Teaching" in the papers published at National Education Congresses in the field of study.

explored. In general, the works found in the CONEDU'S annals show that the use of experimentation in science teaching has mostly been carried out within chemistry teaching in basic education and published in the event as an experience report.

It is hoped that this research represents a two-way street, in which, at the same time as discussing the predominance of the use of experimentation by research approaches, areas of knowledge and levels of schooling, it provokes the curiosity of teachers to use experimental activities and laboratory classes in their pedagogical actions, with the aim of promoting greater learning in the teaching of natural sciences. Since experimentation can encourage students to do science, to recognise themselves as actors who participate in the structuring of knowledge and, through observation and interaction with experimental materials, to better understand scientific phenomena.

REFERENCES

BARDIN, L. **Content analysis**. São Paulo: Editions 70, 2011.

CARVALHO, H. W. P.; BATISTA, A. P. L.; RIBEIRO, C. M. Teaching chemistry from a dynamic-interactive perspective. **Experiências em Ensino de Ciências**, v. 2, n. 3, p. 34-47, 2007.

GUIMARÃES, C. C. Experimentação no Ensino de Química: Caminhos e Descaminhos Rumo à Aprendizagem Significativa. **Química Nova na Escola**, v. 31, n. 3, p. 198- 202, 2009.

MAYER, S. F. **Methodological innovation in the classroom with the use of Concept Maps in higher education.** 2013. 97 f. Dissertation (Master's in Science Teaching), University of São Paulo, SP, 2013.

RAUSCH, R. B. Teacher-researcher: conceptions and practices of teachers working in basic education. **Revista Diálogo Educacional**, , v. 12, n. 37, p. 701-717, 2012.

SÉRE, M. G.; COELHO, M. S.; NUNES, A. D. Papel da experimentação no ensino de física. **Caderno Brasileiro Ensino Física**, v. 20, n.1: 30-42, 2003.

SILVA, V. G. **The importance of experimentation in chemistry and science teaching.** 2016. 42f. Monograph (Degree in Chemistry), Universidade Estadual Paulista, Bauru, SP, 2016.

SUÁREZ, D. H. The narrative documentation of pedagogical experiences as a research-action-training strategy for teachers. In: PASSEGGI, M. C.; BARBOSA, T. M. N. **Narrativas de formação e saberes biográficos**. Natal: EDUFRN, 2008.

OLIVEIRA, M. C. A.; TRINDADE, G. S. Analysis of articles presented at the National Biology Teaching Meetings (ENEBIO) on the theme of practical experimental classes. In: National Meeting of Research in Science Education, 9, 2013. **Proceedings...** Águas de Lindóia: IX ENPEC, 2013.

Buy your books fast and straightforward online - at one of world's fastest growing online book stores! Environmentally sound due to Print-on-Demand technologies.

Buy your books online at
www.morebooks.shop

Kaufen Sie Ihre Bücher schnell und unkompliziert online – auf einer der am schnellsten wachsenden Buchhandelsplattformen weltweit! Dank Print-On-Demand umwelt- und ressourcenschonend produzi ert.

Bücher schneller online kaufen
www.morebooks.shop

FSC
www.fsc.org
MIX
Papier aus verantwortungsvollen Quellen
Paper from responsible sources
FSC® C105338